IMAGES
of America

COMMERCIAL FISHING ON THE OUTER BANKS

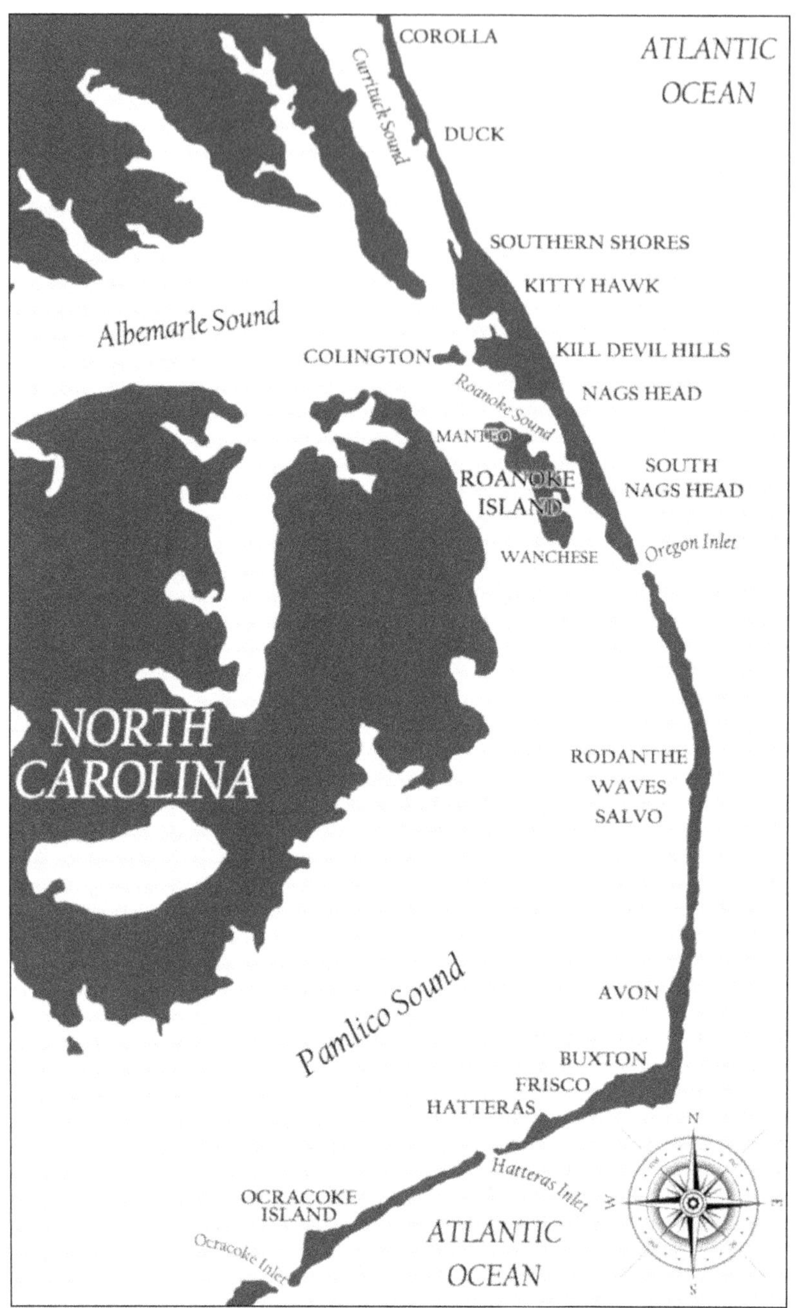

The Outer Banks of North Carolina was isolated from much of the rest of the country until 1928, when the first bridge was built to the island strand. The inhabitants were a resourceful, resilient people who worked side by side to scrape together the necessities of life. The endless offerings of the sea provided the bulk of their diet and a way to earn cash. (Courtesy of Coastal Impressions.)

On the Cover: Lowering a 300-pound block of ice into a shad boat at Mill Landing in Wanchese was no easy task for those on the dock or those on the water. Having access to ice turned fishing from a subsistence activity to a moneymaking enterprise when fishermen had a way to keep their catch fresh and viable for market. (Courtesy of the State Archive of North Carolina.)

IMAGES of America

COMMERCIAL FISHING ON THE OUTER BANKS

R. Wayne Gray and Nancy Beach Gray

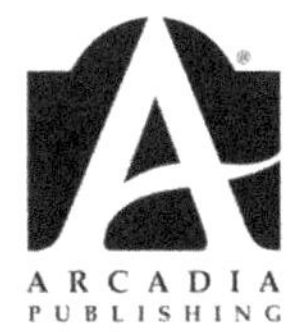

ARCADIA PUBLISHING

ISBN 978-1-4671-0335-0

Published by Arcadia Publishing
Charleston, South Carolina

Library of Congress Control Number: 2018961857

For all general information, please contact Arcadia Publishing:
Telephone 843-853-2070
Fax 843-853-0044
E-mail sales@arcadiapublishing.com
For customer service and orders:
Toll-Free 1-888-313-2665

Visit us on the Internet at www.arcadiapublishing.com

To Willie Etheridge Jr., whose reputation as the finest Outer Banks commercial fisherman is unsurpassed

Contents

Acknowledgments

No book is a complete product of its author or authors. We are grateful to a lot of hardworking, knowledgeable people who gave their time to answer our many questions, share histories, and provide rare photographs about a subject they were passionate about.

For their generous support in the writing of this book, we are indebted to Caitrin Cunningham and Arcadia Publishing. Caitrin, our editor on two previous books, advised and guided us in many ways. We are blessed to have such a valuable advocate.

We are grateful to Benny O'Neal, Frog Tillett, Willie Etheridge III, Colby O'Neal, Ashley O'Neal, Willy Phillips, Moon Tillett, Wayland Baum, and Nicole Harper for talking with us about the commercial fishing industry.

We had the good fortune to interview Billy Carl Tillett, who opened up his files to us on several occasions; Mark Vrablic; Tracy Payne; Joe Wilson; Dave Watkins; Faron Daniels; and Joey Daniels Jr.

We received great support from Philip and Amy Howard and Alton Ballance, all lifelong Ocracokers we count as friends.

We are also grateful for the help of Bob Doxsee; James G. Meade, PhD; Robin Holt; Laura Domingue; Alvah Ward; Lish Meekins; Britton Shackelford; Jeanette Ambrose; and Lou Ellen Quinn.

The staff at the Outer Banks History Center, the Dare County Library, the North Carolina State Archives, the National Park Service, and the Smithsonian Institution contributed tremendously.

We credited all the photographs we used by giving acclaim to a brilliant photographer or by acknowledging a generous contributor.

Many images in this volume appear courtesy of coauthor Richard Wayne Gray (RWG), the Outer Banks History Center (OBHC), the State Archive of North Carolina (SANC), and the National Park Service, Cape Hatteras National Seashore (NPS, CHNS).

INTRODUCTION

The Outer Banks stretch along most of North Carolina's coastline. They have been described as a string of constantly changing sandbars separating the Atlantic Ocean from the mainland. These narrow barrier islands act as protective buffers from nor'easters and storms for the inset islands and mainland.

The Algonquin Indians inhabited this region well over a thousand years ago, drawn to the area for its seasonal bounty of seafood, as is evidenced by many of John White's drawings, as well as large mounds of shells still in existence in some places on the Outer Banks today. Seafood made up a big part of the Algonquins' diet.

The Algonquin weir, which is very similar to today's pound net, was copied by the first settlers and used extensively during historic shad runs.

Subsistence fishing with the Algonquins carried over to the early settlers, a hearty clan who migrated to the Outer Banks for many reasons. Fishing was one of them. Some were shipwrecked, of course, and others were outlaws, but many of them were Virginia planters who were drawn by cheap land, mild winters, and open range for cattle. A subsistent society, the people were bound together by fishing, and later, they commercially fished for profit.

The first recorded occurrence of Outer Bankers making earnings from the sea was about 1665 when the first inhabitants of Colington, one of the islands protected by the barrier strand, shipped 80 barrels of oil processed from drift whales that came ashore. Later, Peter Carteret, whose uncle was one of the Lords Proprietors, kept records showing that he sent hundreds of barrels of whale oil to London.

Another interesting but shameful fishery took place near Ocracoke Inlet at a location called Shell Castle Island in 1790. A porpoise fishery existed where thousands of bottlenose dolphins were brutally killed for their oil and hides. This industry continued on Hatteras Island for over 100 years. Men who worked in the factory said they despised the work, but they had to make a living for their families.

From the time of the early settlers to today, Outer Bankers have always looked to the sea to provide seafood and other natural resources as a crucial part of their sustenance and economy. The local communities were built largely on whatever was taken from the surrounding waters: the ocean, sounds, and tributaries. By its very nature, fishing is a struggle—a fight between man and nature. Yet, the rewards are obtainable and resources from the sea have always benefited the Outer Bankers. The variety and uses seem endless, from processing menhaden into fertilizer to capturing diamondback terrapins to angling over shipwrecks for triggerfish. Mainstay fisheries like crabs, shrimp, and oysters have endured for years through many clashes, like the famous oyster wars when oystermen from Chesapeake Bay and other areas came with sizable boats and dredges and almost wiped out the oysters in the Pamlico Sound.

Large-scale commercial fishing, however, did not begin until after the time of the Civil War. Fisheries before that era consisted mostly of salted fish, like mullet, shad, and herring, which were shipped to Northern markets. These early catches of fish were preserved with salt or smoke because the remoteness of the Outer Banks and the lack of sufficient ice prevented the delivery of fresh seafood to distant places. Clams and oysters were often hawked from horse and cart from house to house for a penny apiece. Salted and smoked fish were traded to communities in Hyde County, across the sound, for bushels of corn and other produce.

In the 1880s, with the advent of the first ice plant at Skyco on Roanoke Island and the big shad run, commercial fishing reached a new level. Newspapers of the time labeled the Outer Banks the "Shad Fish Capital of the World." Fishing began to play a major role in the lives of many men who were suddenly able to make a good living. St. Clair Pugh, of Wanchese, for example, had multiple shad fishing crews and made enough money in one season to build a large mercantile store and a beautiful Victorian home that still stands today. Later, the building of roads and bridges further enhanced commercial fishing. All kinds of seafood could be shipped to places like Baltimore, Philadelphia, and New York.

In time, the traditional fishing industry made a change from sound fishing to ocean fishing. Because of the deepening of Oregon Inlet, ocean trawlers could, for the first time, go back and forth to outlying fishing grounds. Malcolm Daniels and his father-in-law, Willie Etheridge Sr., were the early pioneers of this movement. They converted an old fireboat, the *Faith Evelyn,* and rigged it for ocean trawling. Its captain was Willie Etheridge Jr., the best fisherman in the area. Other boats soon entered the contest when fishermen saw their success. Shrimp boats and retired World War I and World War II vessels, like subchasers, air-sea rescue boats, and fireboats, were quickly and cheaply transformed into offshore fishing vessels. As fish landings increased, large steel trawlers, many of them made in Wanchese, joined the fishing fleet, and Wanchese soon became one of the major seafood handling ports on the East Coast.

It would not be long, however, before Oregon Inlet, because of shoaling, became known as a dangerous inlet. Because it required and lacked proper dredging, safe passage was no longer possible and fishing activities decreased. Oregon Inlet, the key to big fisheries, was unstable, and the decades-long push for a jetty at Oregon Inlet has come to an impasse.

Today, despite the shoaling of Oregon Inlet and many state and federal rules and regulations regarding fishing, several Outer Banks communities still rely on commercial fishing. The small village of Wanchese, the hub of seafood transportation, ships about 90 percent of the seafood landed on the Outer Banks—from squid to sushi-grade bluefin tuna—to places as far away as China and Japan.

Nonetheless, the survival of the fishing industry is threatened as never before on many sides. Fishermen have to learn to be versatile by using different gear and different boats to fish for a variety of species. They see a need for a national voice in order to exist. Cultural, political, and economic problems plague the commercial fishing industry, but it is a good bet that fishermen will win in the long run. Family livelihoods and businesses are an important part of our economy and local history. Commercial fishermen, acutely aware of this, will continue to develop new markets and fight for their livelihoods.

One

From Prolific to Protected

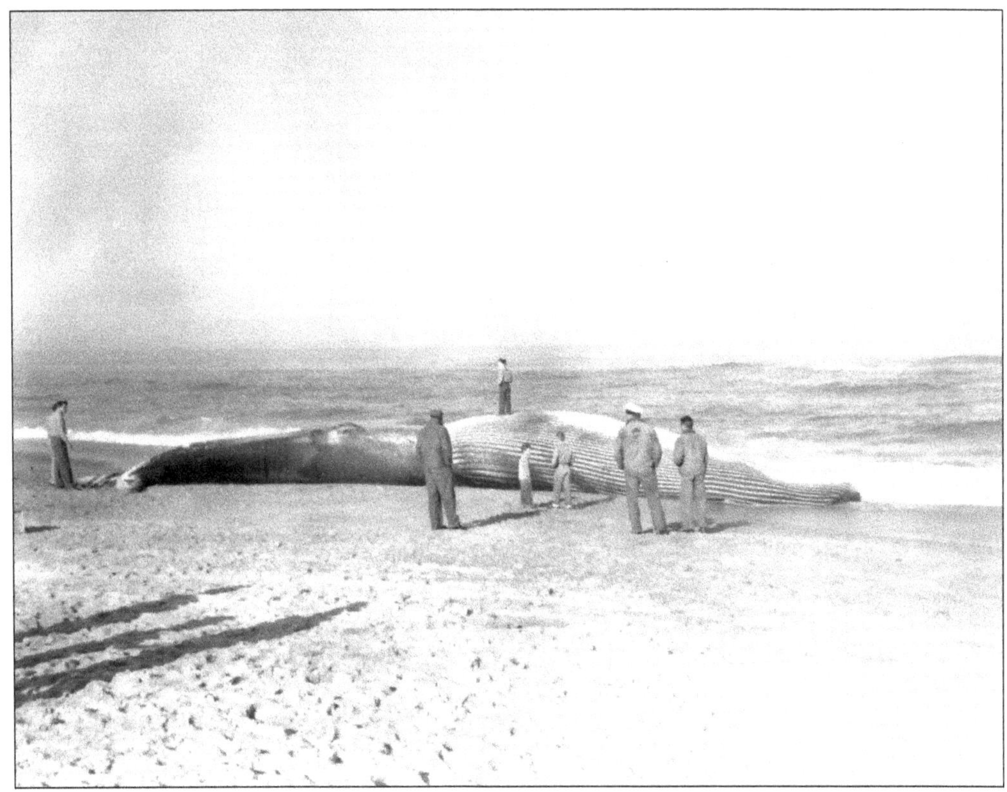

Outer Bankers have always made a living from the sea, by sheer intentional will or by taking advantage of uninvited but welcomed opportunities. As recorded in the 1660s, the first settlers living on Colington Island sold 80 barrels of oil rendered from dead whales, called drift whales, like this whale that washed up near Southern Shores beach in 1955. (Courtesy of NPS, CHNS.)

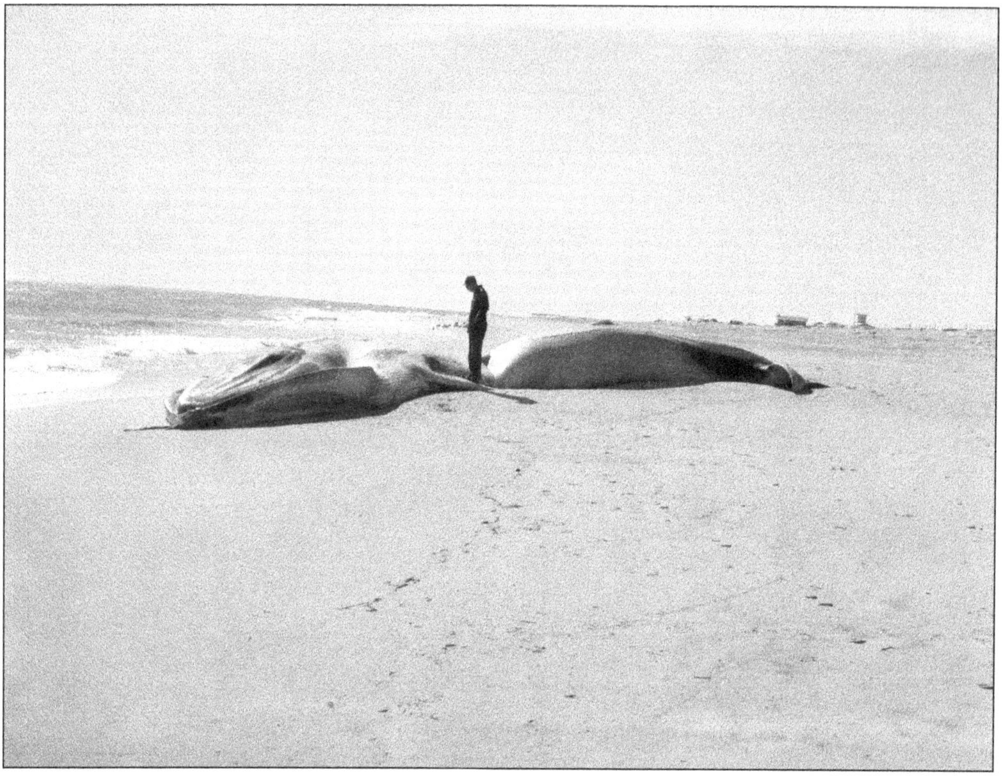

Families would come together to process drift whales. Neighbors would set up portable, makeshift operations to cut up and cook down the blubber in iron pots over wood fires. It could take up to 10 smelly days to render all of the fat into precious whale oil. The whale in this 1971 photograph, taken on the Cape Hatteras National Seashore, was put to no such use. (Courtesy of NPS, CHNS.)

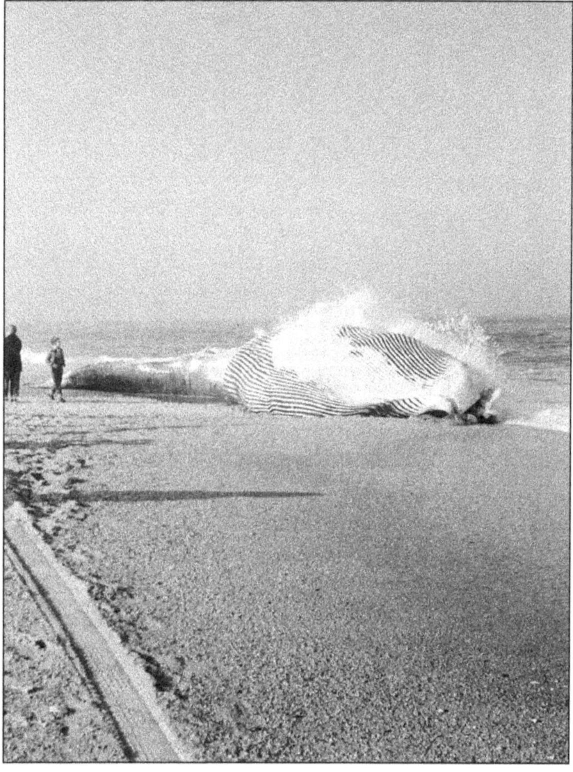

Twice yearly, whales migrate along a specific route just off of the Outer Banks. This whale's life ended on Portsmouth Island in the 1940s. Although New England whalers used Cape Lookout, North Carolina, as a base of operations, there is no written or oral record that deliberate whaling took place on the northern Outer Banks. (Courtesy of Aycock Brown Collection, OBHC.)

It is possible that enterprising local fishermen learned and used whaling methods developed elsewhere. Tall dunes or a tower were needed for the lookout to spot the passing mammals. A boat would be launched into the pod so that a harpooner could hook onto and exhaust a whale. This pilot whale washed up near the old Coast Guard station in Kitty Hawk in 1954. (Courtesy of Aycock Brown Collection, OBHC.)

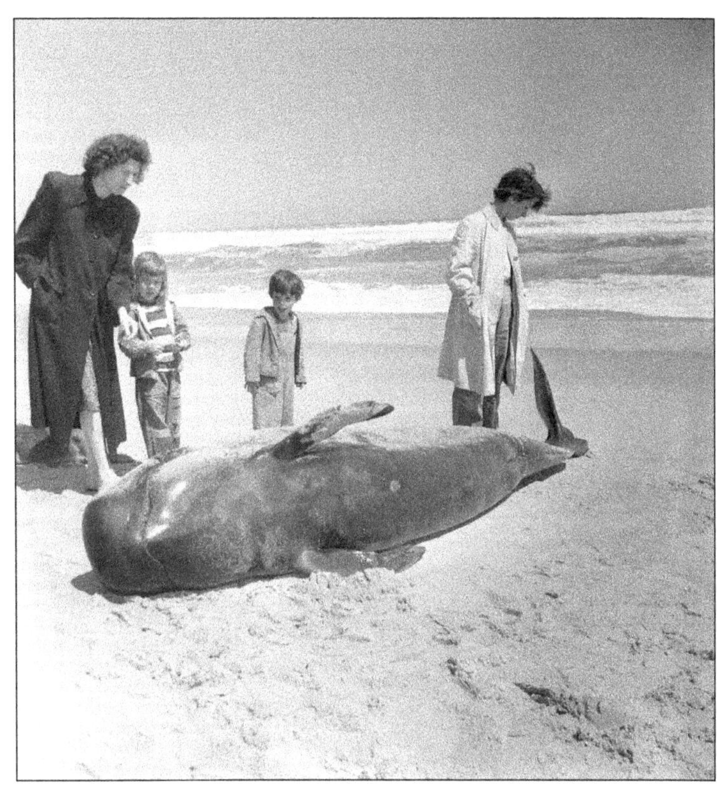

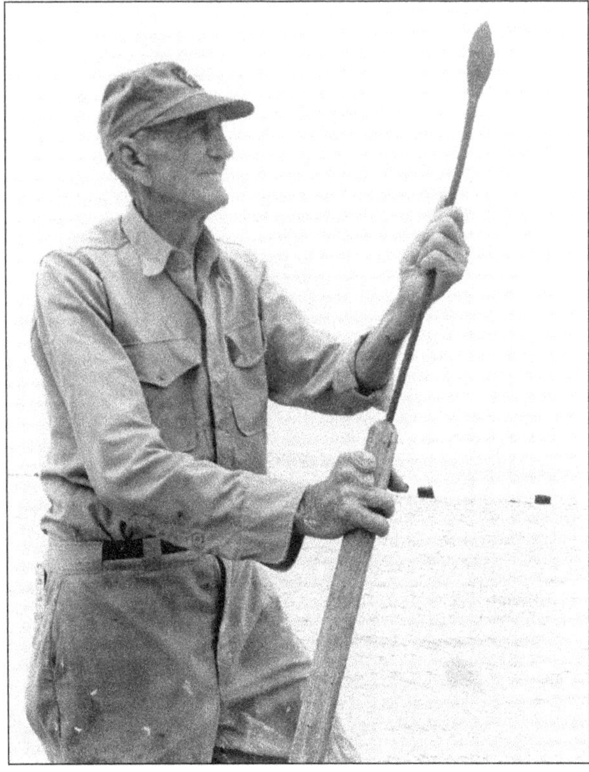

This man from the Cape Lookout area posed with a whaling spear. Whalers were at a great disadvantage in the ultimate battle between man and beast. Six men in an open wooden boat with only handheld harpoons and coils of rope were pitted against a huge, intelligent creature. By the look of this man's right hand, he also had an unforgiving encounter of some kind. (Courtesy of Bruce Roberts Collection, OBHC.)

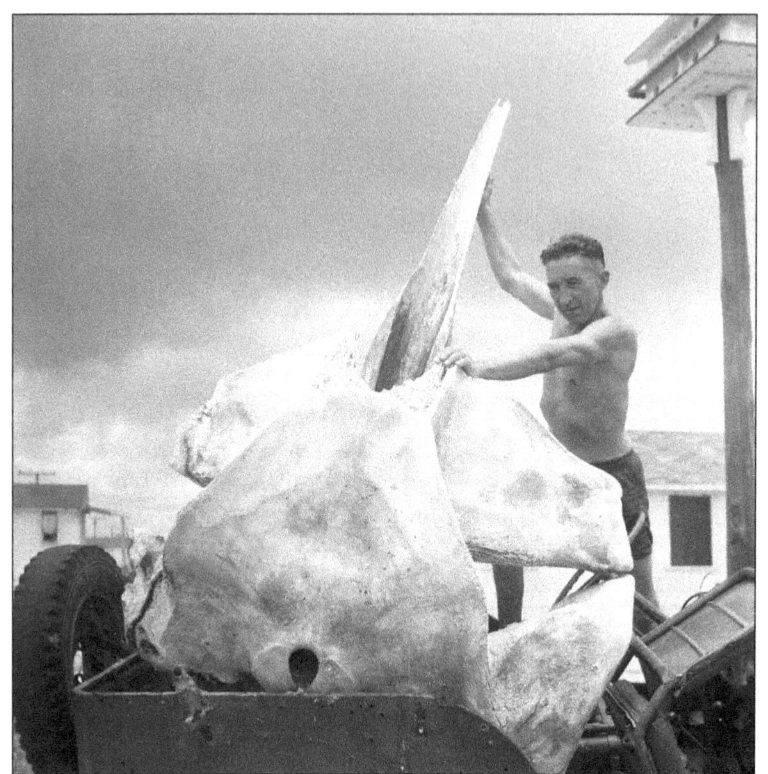

Before the use of petroleum, whale products were essential. Lighthouses used oil to keep lanterns burning. Spermaceti, the highest-quality oil extracted from sperm whale heads, lubricated fine instruments. Baleen, the bonelike feeding filters, was used for corsets and umbrella ribs. This whale skull was loaded in a jeep at Caffey's Inlet in Kitty Hawk around 1953. (Courtesy of Aycock Brown Collection, OBHC.)

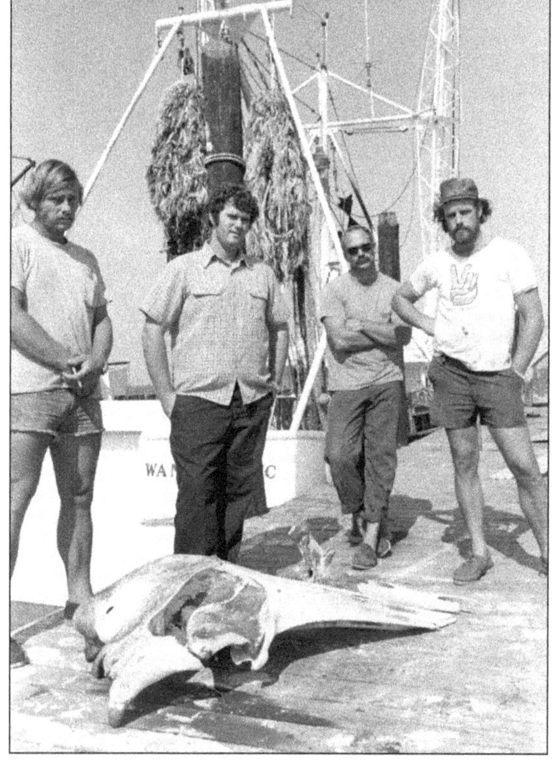

In the 1980s, from left to right, Doug Bateman, Capt. Billy Carl Tillett, John Arendts, and Vernon Lewark pose with a whale skull and vertebrae that were brought up in a net on the *Linda Gayle*. Locals used to love to show off whalebones. Whalebone Junction in Nags Head was named for a 72-foot-long whale skull and backbone displayed in front of the Midgett family's gas station. (Courtesy of Billy Carl Tillett, photograph by J. Foster Scott.)

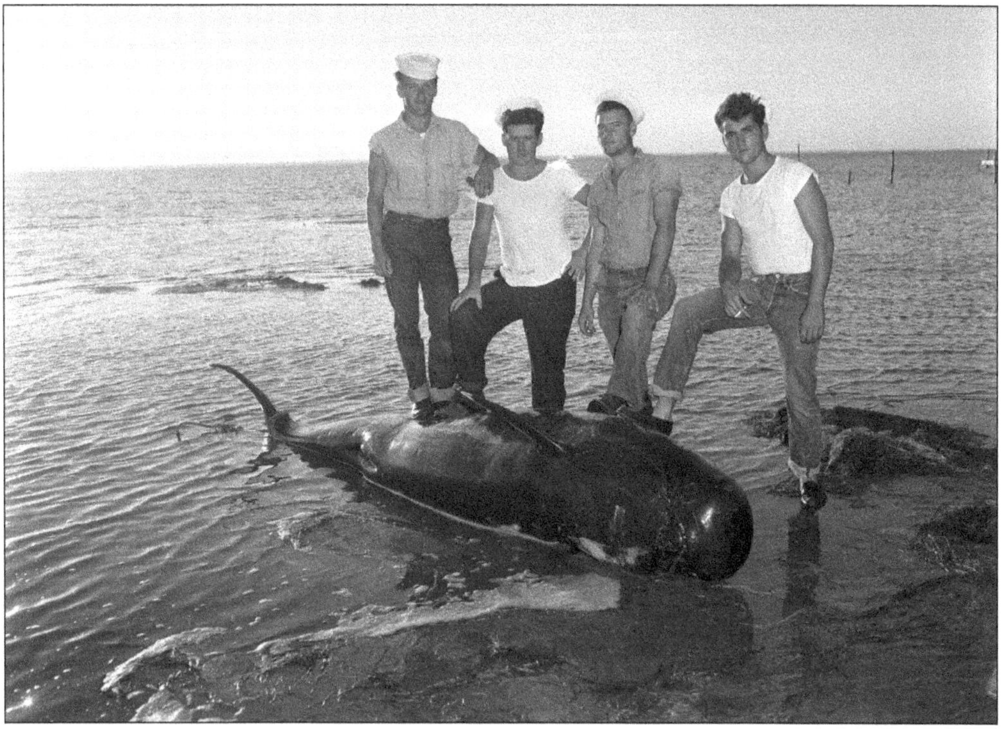

A disoriented pilot whale, perhaps chased by sharks, washed up on Duck's sound side in October 1954. The Navy men stationed at that quiet outpost tried, but failed, to save the whale's life. Nowadays, Duck is a premier vacation destination, but in the 1950s, besides the Navy men, only a few fishing families inhabited the desolate area. (Courtesy of Aycock Brown Collection, OBHC.)

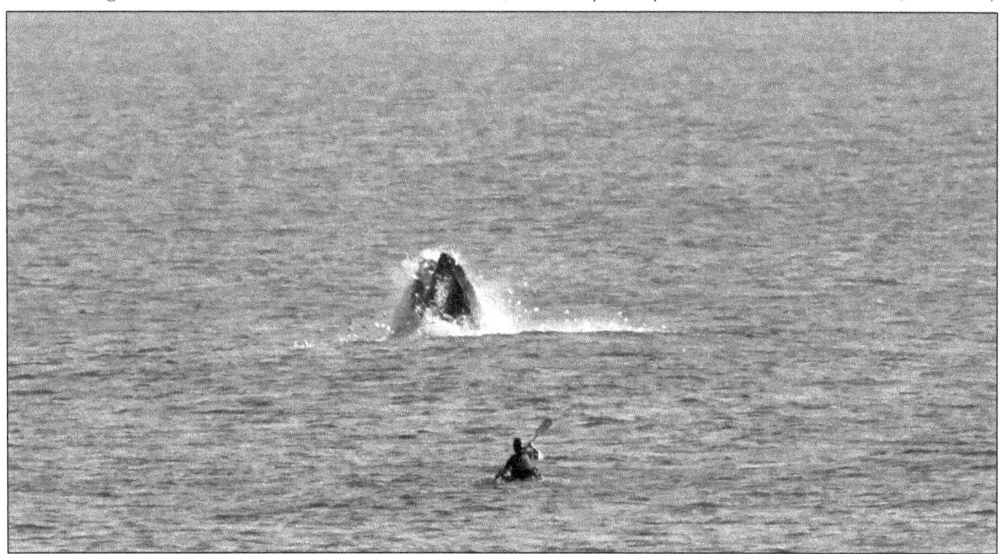

This kayaker must have hoped someone else besides him saw the massive whale breaching near Jennette's Pier in Nags Head. It was most likely a humpback since they tend to swim close to shore. They pass by in December and January on their way to the Caribbean to breed and give birth and, then again, in March and April on their way to polar waters for summer feeding. (Courtesy of Daryl Law.)

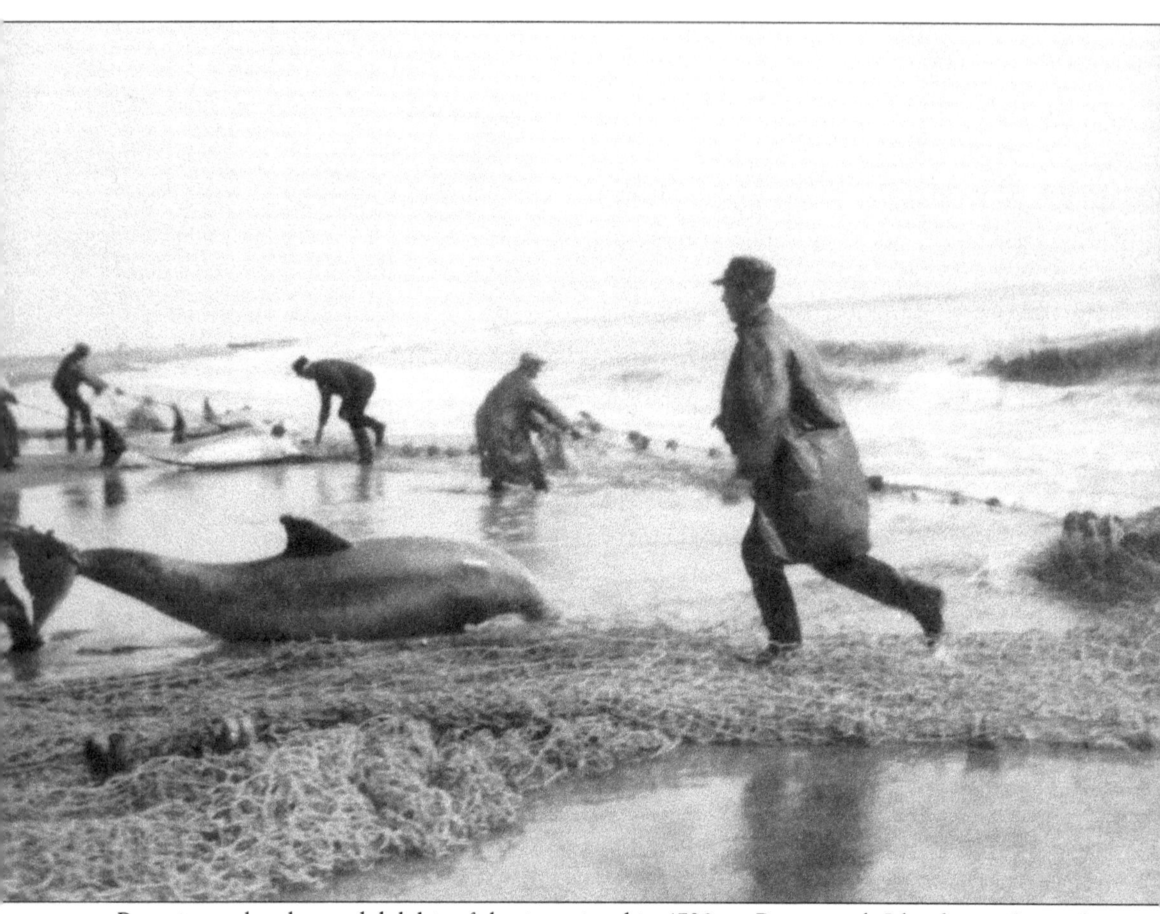

Porpoise, or bottlenosed dolphin, fisheries existed in 1790 on Portsmouth Island near Ocracoke and, later, on Hatteras Island. According to a diary kept by one of the last operators, Hatteras native William Harris Rolinson, the fishermen absolutely detested the work. It was near-freezing cold, but more unsettling, it was a heartless business. Each crew was made up of 14 to 18 men. A "spy" would spot a school and raise a flag to alert men waiting in dories 150 yards offshore. They would quickly row to the beach releasing one end of a 400-yard barrier of 18-inch mesh, heavy-twine net, cutting off the porpoises' forward path. A remaining boat would come to shore after the mammals had passed, thus trapping them in a large semicircle. In this 1928 photograph, Hatterasmen use a sweep-seine net within the outer net to herd the porpoises to shore. (Courtesy of Marine Mammal Program, Division of Mammals, National Museum of Natural History, Smithsonian Institution.)

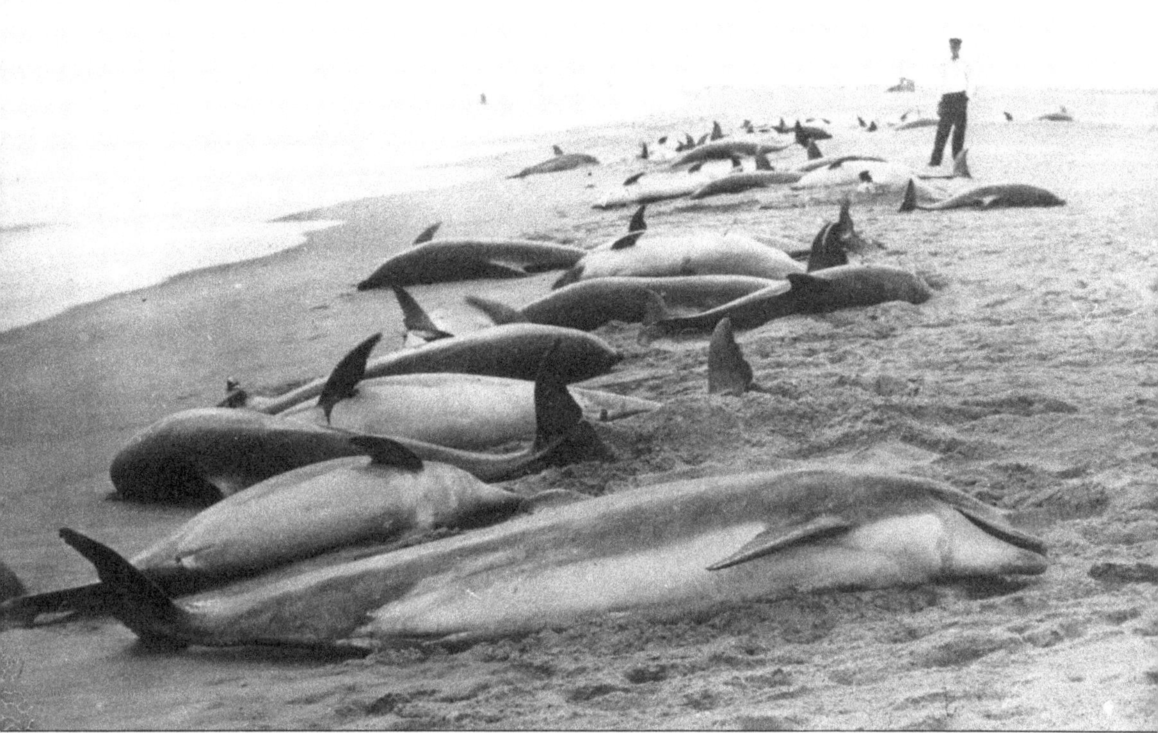

Taking advantage of migration patterns, porpoise fishing started in late December and ended in early April, usually an unproductive time for watermen. The men were grateful for the paycheck, meager though it was. Fishing and processing plant positions were given to local men, but supervisory roles were held by out-of-state entrepreneurs. Three separate processing plants were located sound side on lower Hatteras Island. Although extremely clever, once in the nets, the porpoises seemed resigned to their fate. Lou Angell wrote that the scraping of an oar along the warp lines would keep them from swimming under the net, even though they could have easily done so. If one did happen to jump the cork line, it was difficult to keep the others from following. The fishery lasted on and off from 1790 to 1928, with the heyday being in the late 1800s. (Courtesy of Wilson Special Collections Library, University of North Carolina, Chapel Hill.)

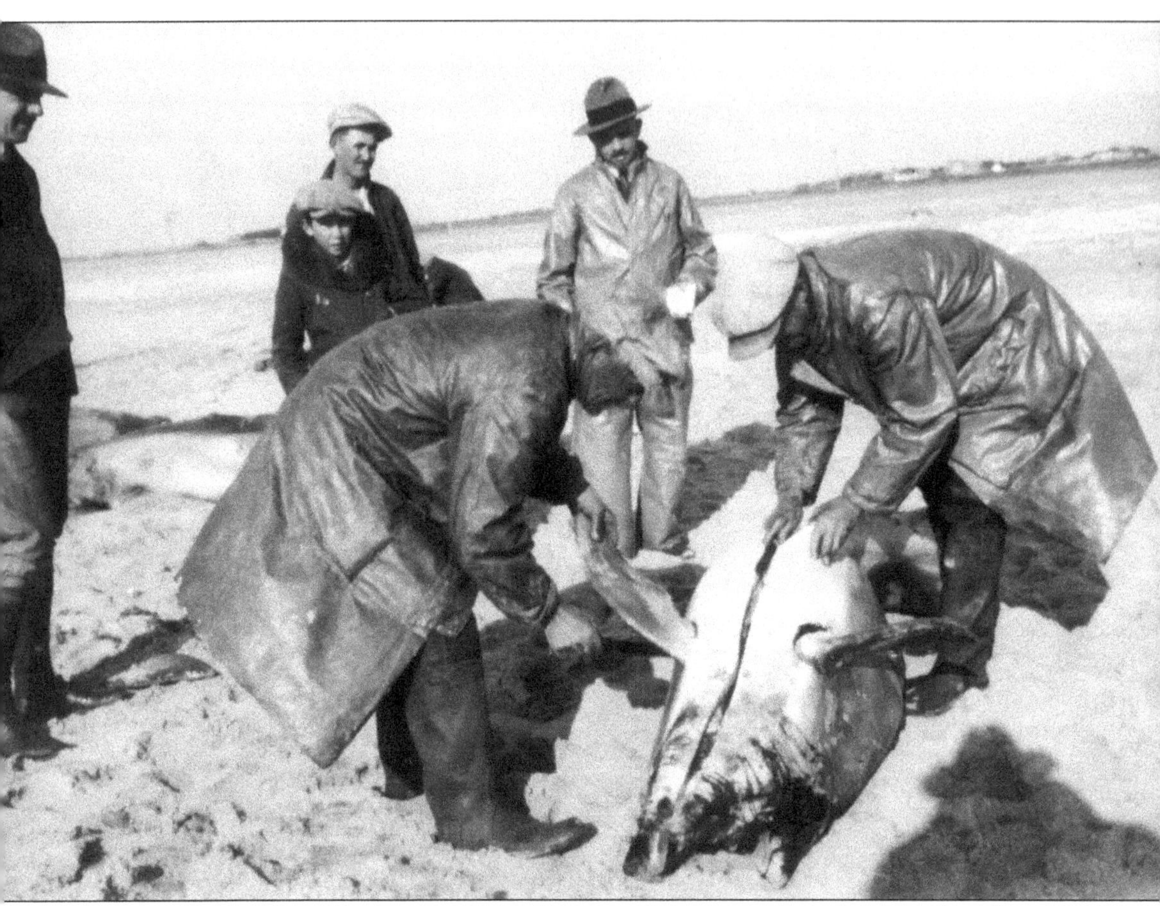

Jarvis Midgett, the tallest young man with a cap, and Augustus Austin, the smaller fellow, watch as a dolphin is skinned on February 10, 1928, the day Dr. Remington Kellogg of the Smithsonian visited Frisco, North Carolina. The melon, a small organ involved in echolocation above the porpoise's jaw, had already been taken for the high-quality oil it yielded. Called Nye oil after the man who perfected it, the lubricant was used in clocks, watches, and other superior mechanisms. Blubber was rendered down for oil, the one-inch-thick hide was used for machine belts, and the carcasses were used for fertilizer. The men hated hooking the porpoises in their blowholes, dragging them ashore, and stabbing them. Even the horses that transported them in carts were afraid of the scene, and the strongest man had to handle the teams. (Courtesy of Marine Mammal Program, Division of Mammals, National Museum of Natural History, Smithsonian Institution.)

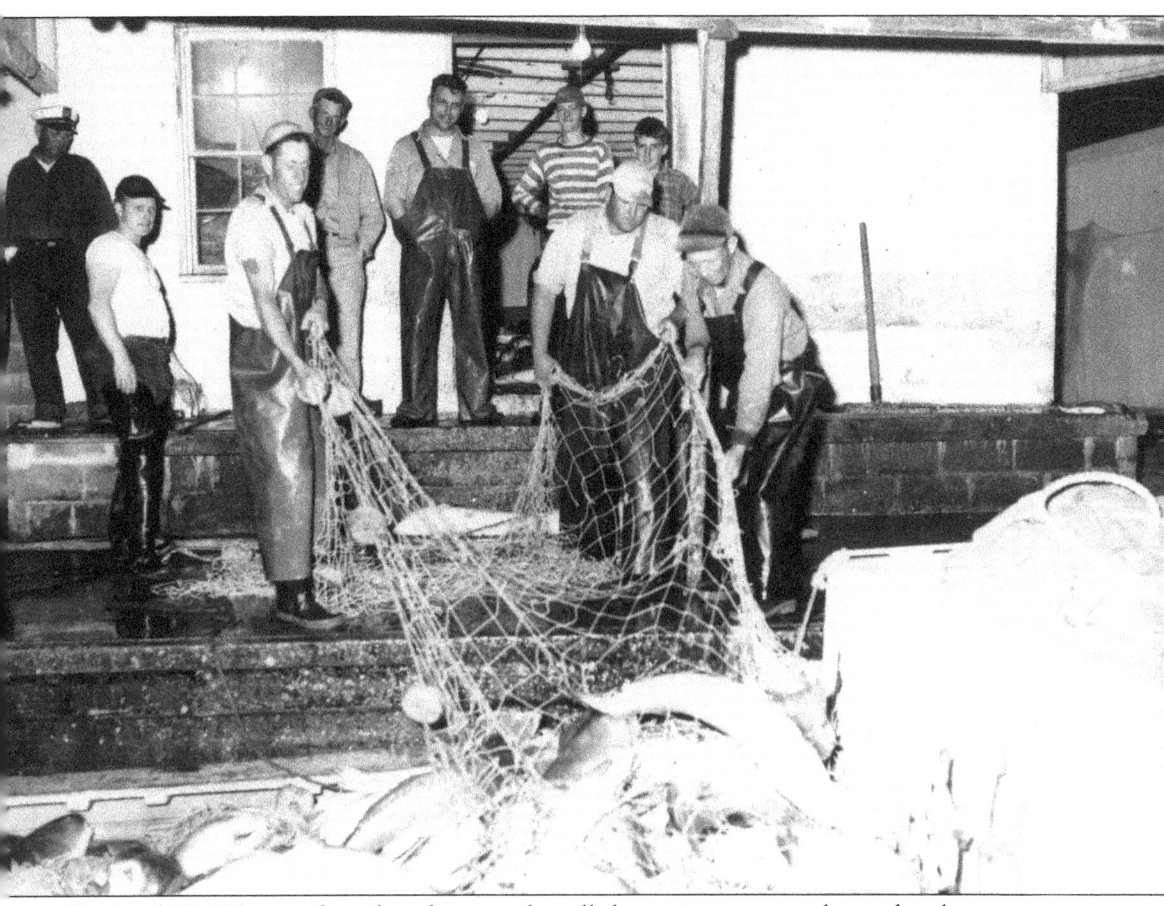

On April 19, 1954, people gathered to see what all the excitement was about when boats came into Wanchese harbor with drum spilling over the sides. It was the result of a springtime run of channel bass at Oregon Inlet. The crews of Willie Etheridge Sr., Dewey Tillett, and Ralph Meekins caught 1,152 drum (plus whatever the men took home with them) on the last set of the day. Fishermen of old described how the water would literally take on a red hue when a school of drum presented themselves as a solid bed racing close to the surface. When sitting on a bench in front of his Nags Head store, Jethro Midgett Sr. claimed he could smell them swimming down the beach before he saw them. Capt. Wayland Baum remembered loading his shad boat with 80 to 90 drum at Oregon Inlet and carrying them to the ice plant at Skyco. Years earlier, his father did the same thing with the same shad boat, except it was under sail power and had to be rowed when there was no wind. (Courtesy of Roger Meekins.)

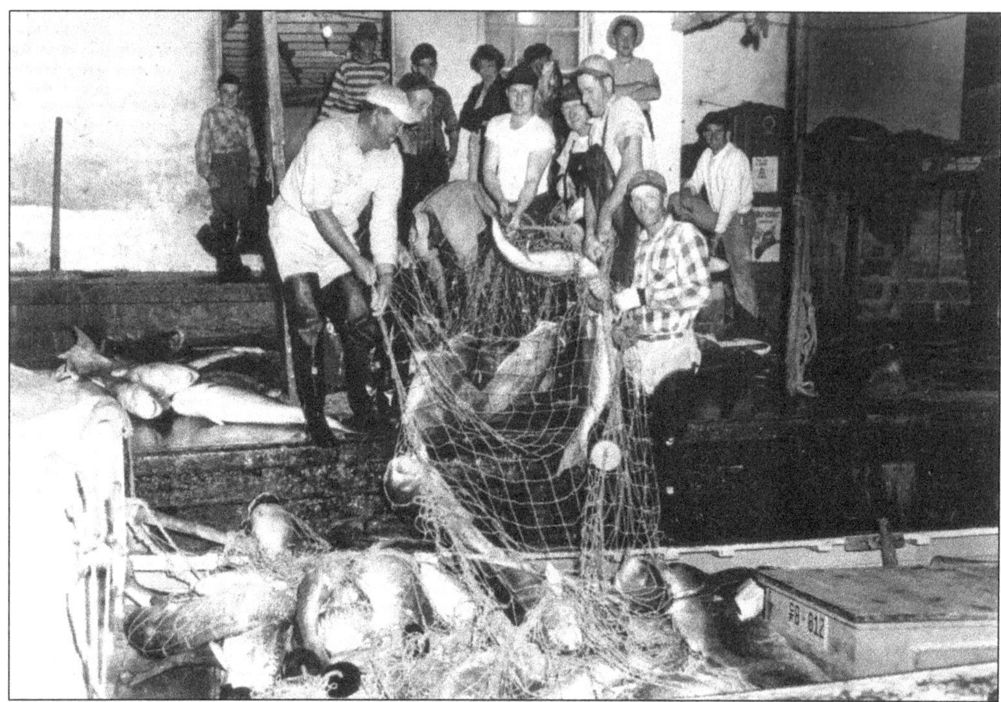

Working by the light of single lightbulbs, the men labored to get the "copper-colored beauties," as the photographer described them, out of nets that were already torn from the stress of such a large haul. The fishermen were paid $1,800 for the lot, and the 35-to-40-pound fish sold for $2 apiece retail. Dewey Tillett predicted it would take a couple of days for them to mend their nets. (Courtesy of Roger Meekins.)

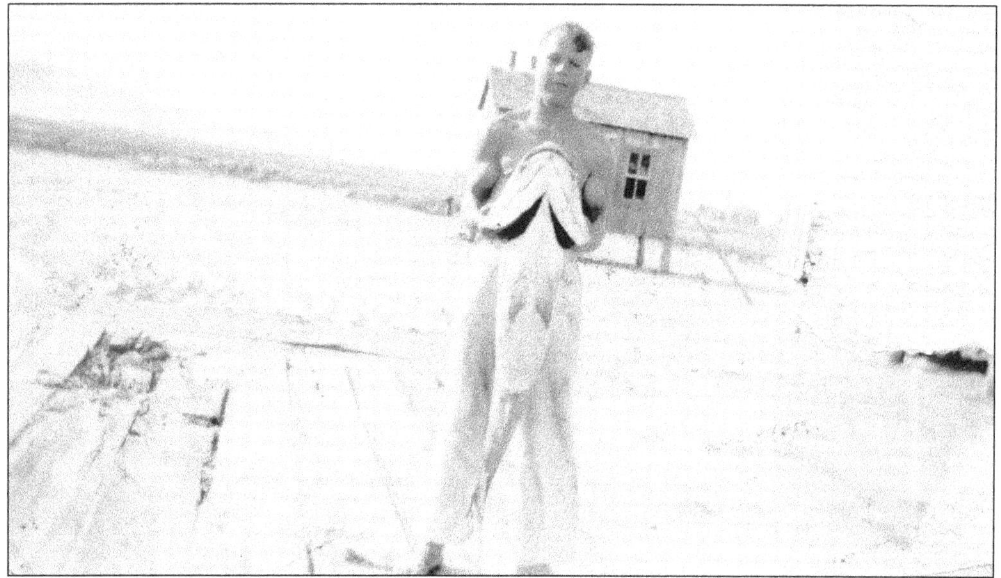

With a faint outline of Jockey's Ridge sand dunes in the background, teenaged Jethro Midgett Jr. showed off a prized drum that was most likely caught in the net of the family's beach seine operation. His parents made a bold step when they moved their Nags Head sound-side store to the ocean side to better serve the growing population of summer vacationers. (Courtesy of Jeffrey G. Midgett.)

Some 30 years later, Jethro Midgett Jr. had not lost his desire to pose with a drum. This time, he was inside Midgett's Seafood fish house, built next to his parents' store. Two drum canneries, one in Wanchese and one in Skyco, existed until the 1940s. Bob Scarborough had wildlife artist, Frank Stick, design the label for his product canned in Wanchese. (Courtesy of Jeffrey G. Midgett.)

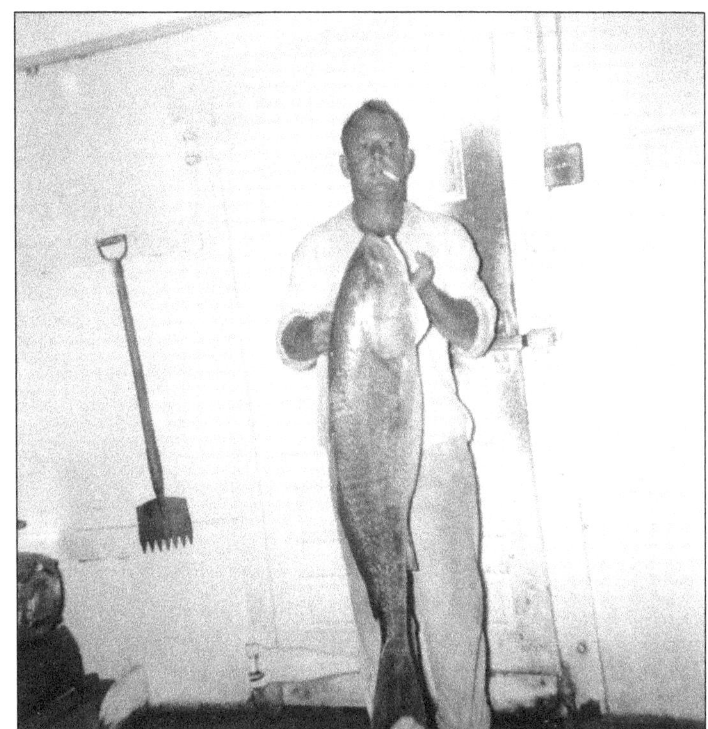

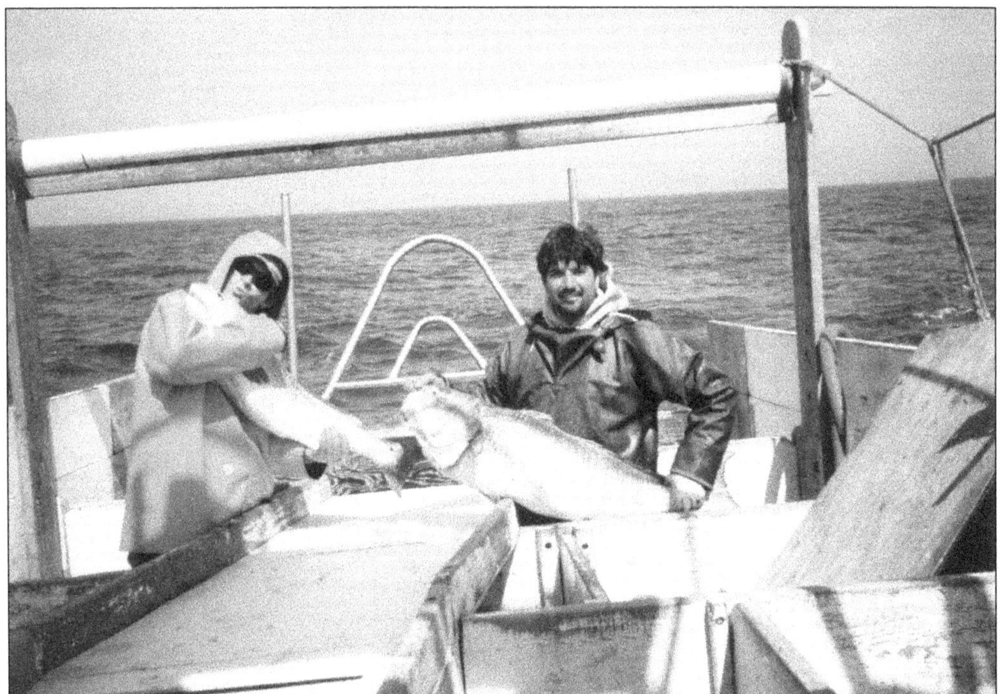

The drum caught by Capt. Britton Shackelford and his mate, Caine Livesay, look good enough to kiss, but since the drum were too big to keep, it was a goodbye kiss. A best-loved dish, the old drum dinner consisted of boiled potatoes and drum topped with raw onion and fried pork cracklings and grease. Fishing families thought this was food fit for a king. (Courtesy of Britton Shackelford.)

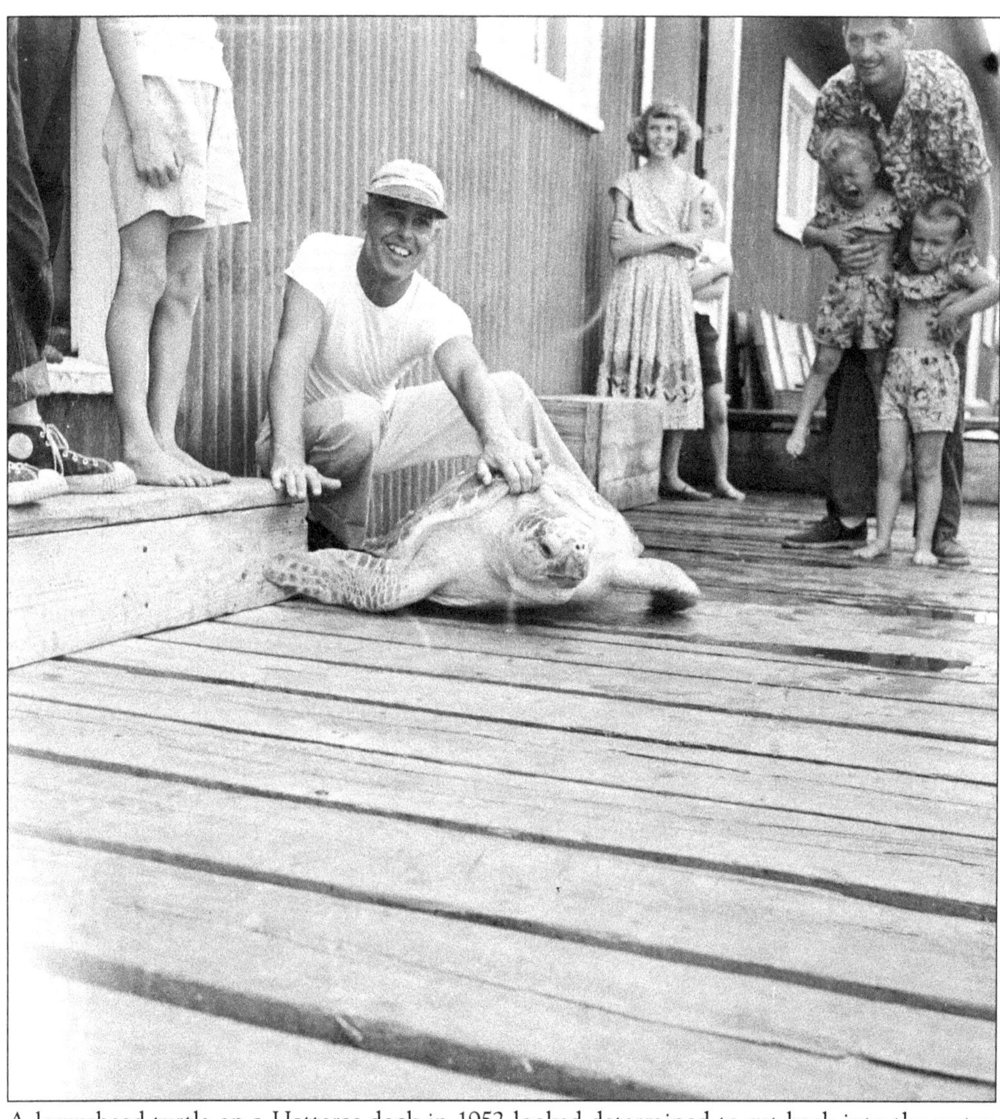

A loggerhead turtle on a Hatteras dock in 1952 looked determined to get back into the water. Before the Endangered Species Act of 1973, sea turtles were caught and their eggs and flesh were eaten. They were usually obtained as a bycatch when fishing for other species. Farther south, near Wilmington, divers plunged into the deep and caught turtles by hand. Outer Bankers ate the firm meat in soup, hash, or fried. Always a delicacy, turtle steaks were sold by one retail seafood market for $10 a pound, a high price in the 1970s. Turtles were eaten in colonial days, too, as recorded by Englishman Charles Johnson, who met a group of watermen slaves near the Alligator River. They were transporting loggerheads, terrapins, and snapping turtles to Columbia, North Carolina, and Johnson speculated that the slaves were conducting trade after work or on the sly. (Courtesy of Aycock Brown Collection, OBHC.)

The trawler *R-Dream* pulled up to the Wanchese docks with a prehistoric-looking creature and caused quite a commotion. In the 1960s, villagers had never seen such a large leatherback turtle. From left to right, Doogie Pledger, Capt. Dennie Daniels, Bernie Daniels, and Rollins Beasley were also mystified by the deep-sea turtle. (Courtesy of Robin Daniels Holt, photograph by Aycock Brown.)

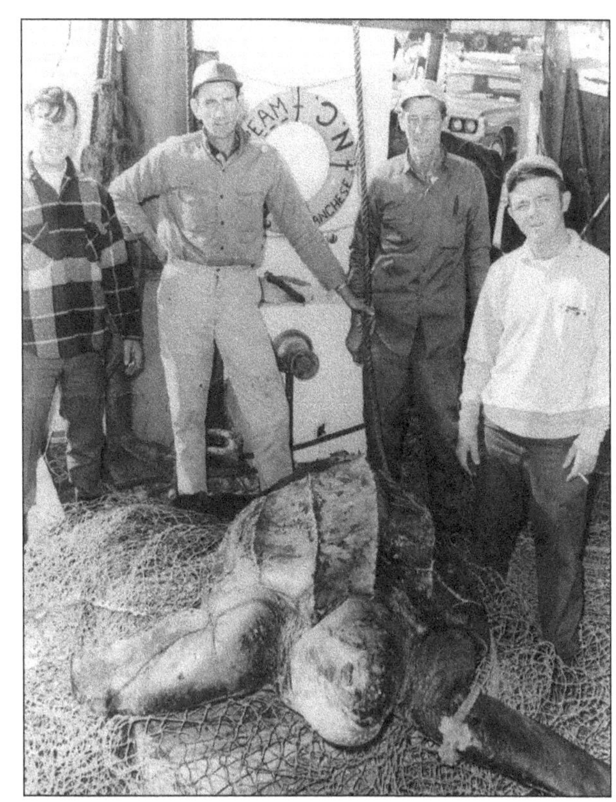

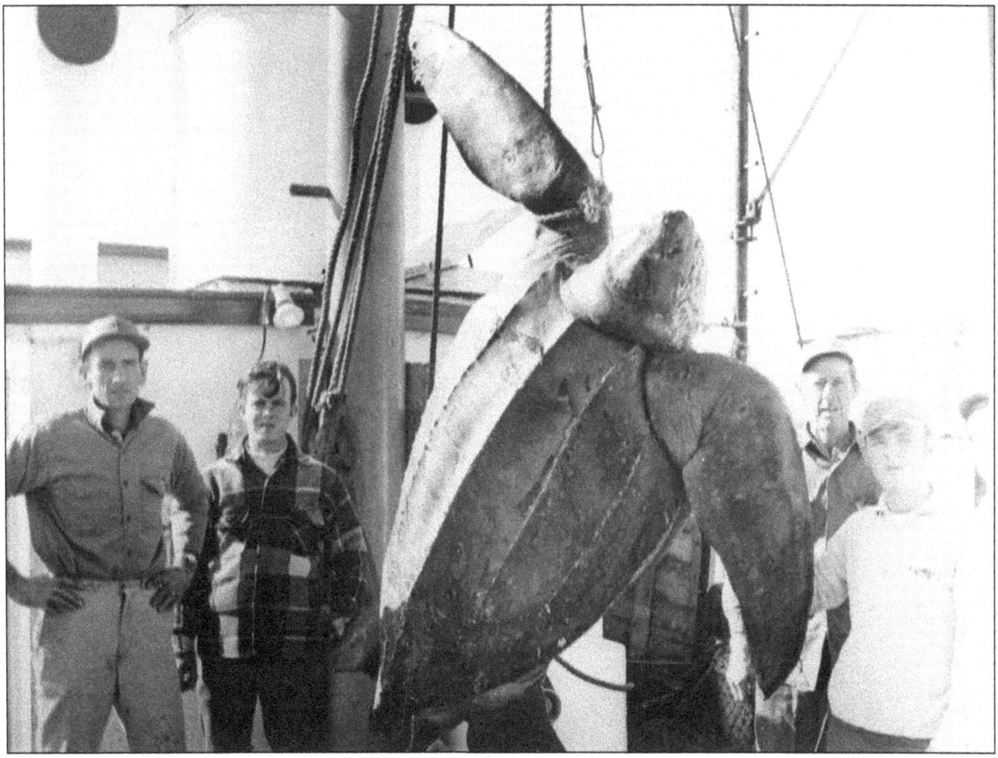

A leatherback turtle hoisted in the air is a horrifying sight by today's standards. By law, modern fishermen rig their nets with turtle excluder devices that allow unintentionally entrapped turtles to escape through chutes before drowning. Even though TEDs have many failures and drawbacks, fishermen are heavily fined if a turtle is caught in their net. (Courtesy of Aycock Brown Collection, OBHC.)

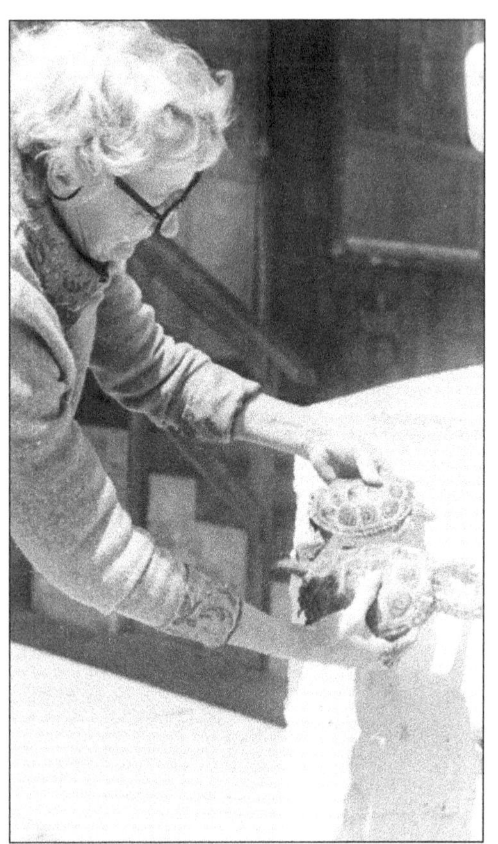

At one time, diamondback terrapins, like the ones held by Nellie Myrtle Pridgen, commanded the highest price per pound of any other product in fisheries. Fancy New York City restaurants served terrapin soup made with heavy cream and sherry. Nellie Myrtle's father kept terrapins in a sound-side pound in Nags Head and dipped them out as needed for food or sale. (Courtesy of Aycock Brown Collection, OBHC.)

Always ready for a fight, snapping turtles wait for one false move from Harry Niser (left) and Jean Domingue (right). Thomas Tillett of Manns Harbor trapped turtles along ditches and canals all summer. He would retain them in a pound, feeding them table scraps and once a truckload of watermelons, until he shipped them live up North in the fall. His tax return stated that he made $84.24 in turtles in 1968. (Courtesy of RWG.)

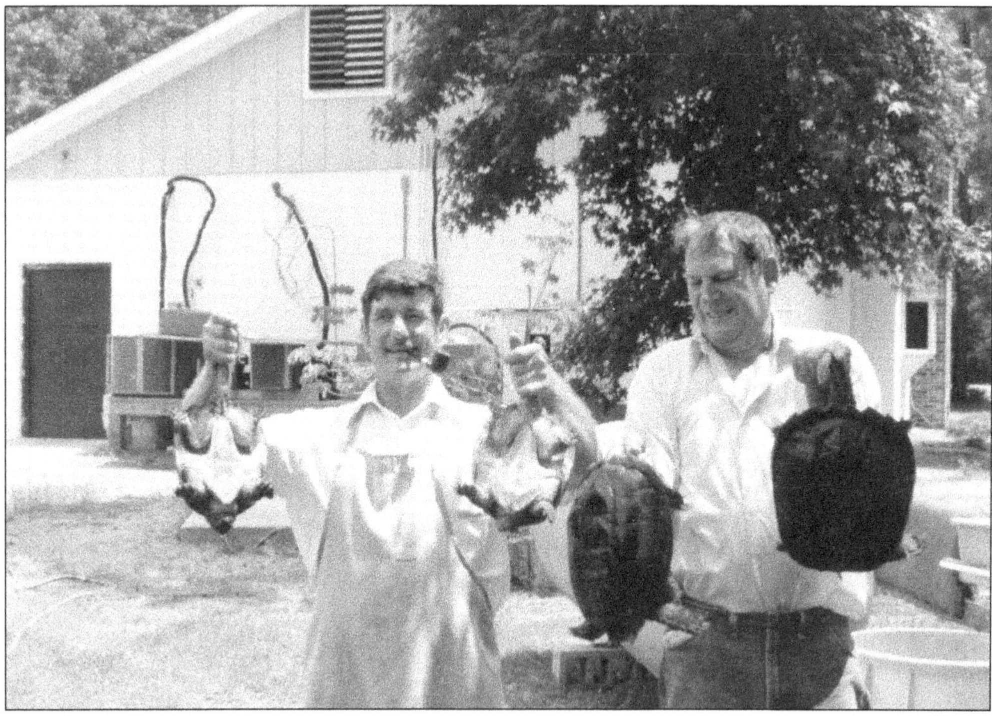

When it was still legal, Billy Carl Tillett (left) and Ralph Wayne Johnson caught this 545-pound, nine-foot-long sturgeon while trawling. It was sent to a dealer in New York, but whether it was too big to handle or because it was caught in the fall with no valuable caviar, they never received a paycheck. They usually got $2 to $3 a pound for the 50-to-60-pound sized sturgeons. (Courtesy of Billy Carl Tillett.)

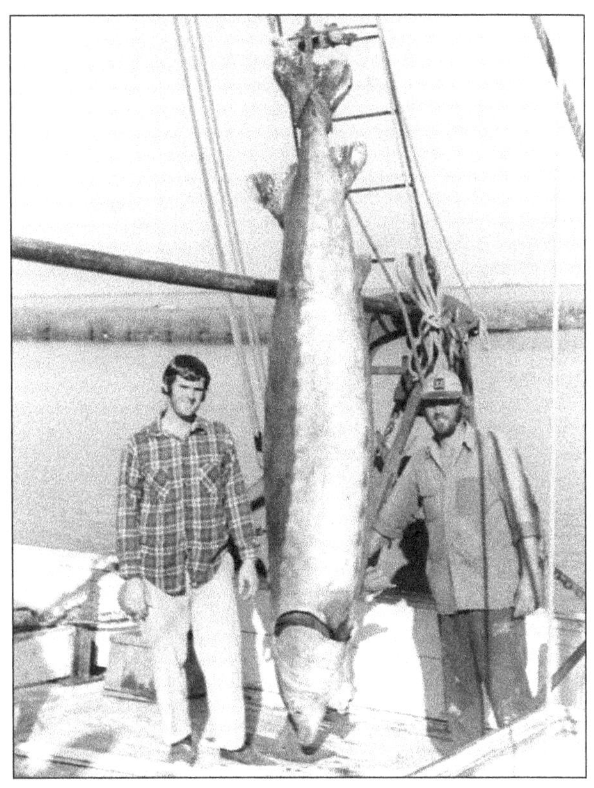

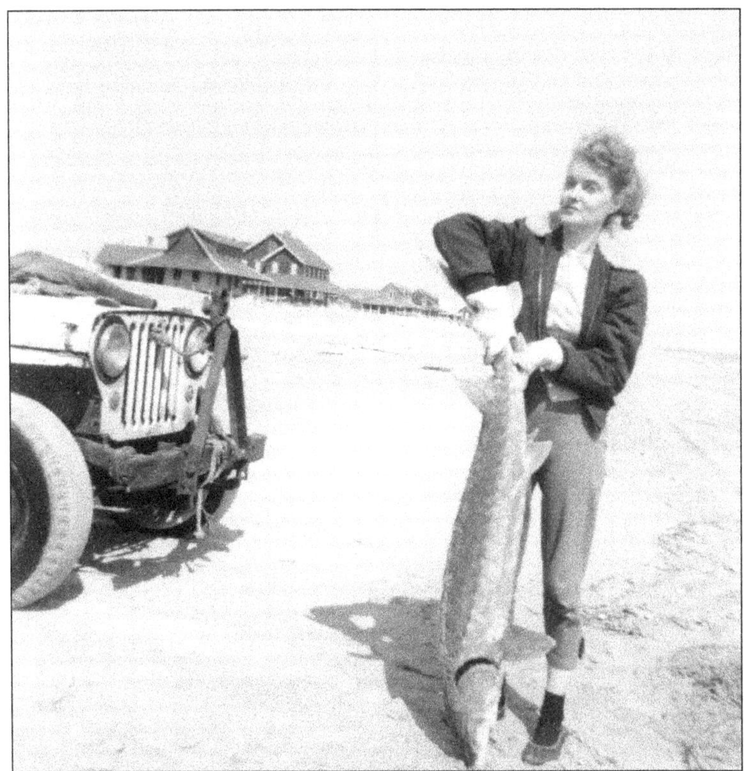

Cecelia "Beauty" Midgett was so petite that she did not weigh much more than the sturgeon caught in her husband's net in 1960. The World War II surplus jeep with four-wheel drive was used to pull the net onto the wide beach in front of the Unpainted Aristocracy, a row of historic cottages in Nags Head. Sturgeons are anadromous, meaning they migrate from saltwater to spawn in freshwater. (Courtesy of Jeffrey G. Midgett.)

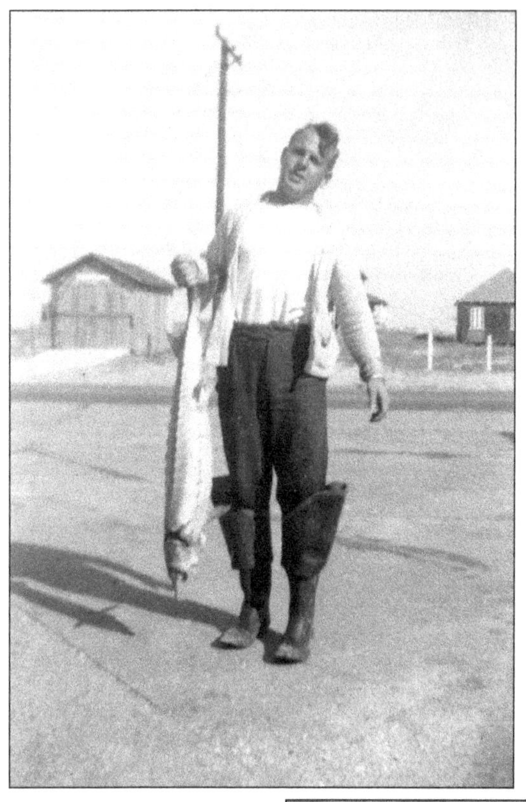

The characteristics of sturgeon have not changed since they were entombed as fossils. Under the bony plates of armor is a layer of thick orange fat, but the payday is in the caviar. The eggs in the sturgeon held by Jethro Midgett Jr. will be worked by his father through a screen and then salted. The caviar sold for $7 a pound in the 1960s. (Courtesy of Jeffrey G. Midgett.)

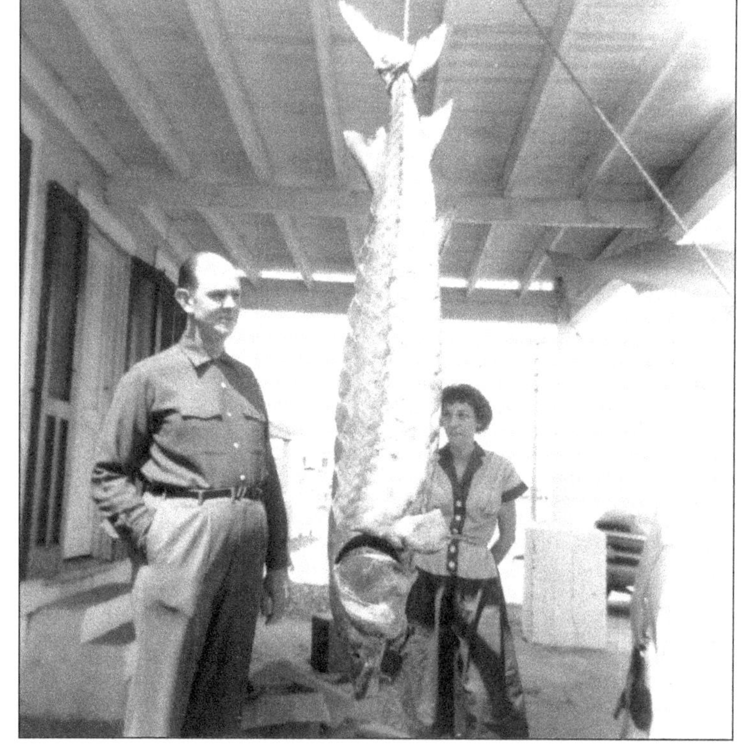

Dr. John Highsmith of Plymouth, North Carolina, admired an Atlantic sturgeon in 1956. Coauthor Wayne Gray dropped out of college one spring semester to sturgeon fish with his uncle. They set a series of nets with dories about a mile from the shore in front of Jockey's Ridge. After that hard, cold spring of catching and skinning the monsters, Gray was motivated to get back to school. (Courtesy of Jeffrey G. Midgett.)

Two

HERITAGE FISHERIES

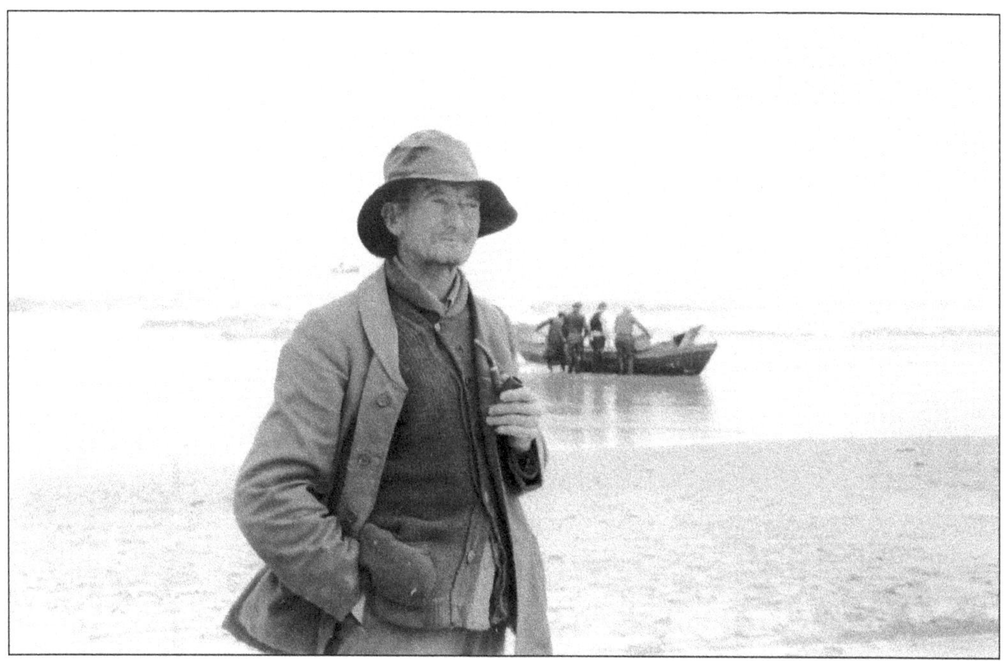

This fisherman struck the quintessential pose of an old salt for this January 1939 photograph, taken north of Duck. The crew busily handled his boat, while a similar sailing vessel plied ocean waters within sight of the shore. Most local fishermen did not venture out into the Gulf Stream until after World War II and the development of Loran, a radio transmitter navigational system. (Courtesy of NPS, CHNS.)

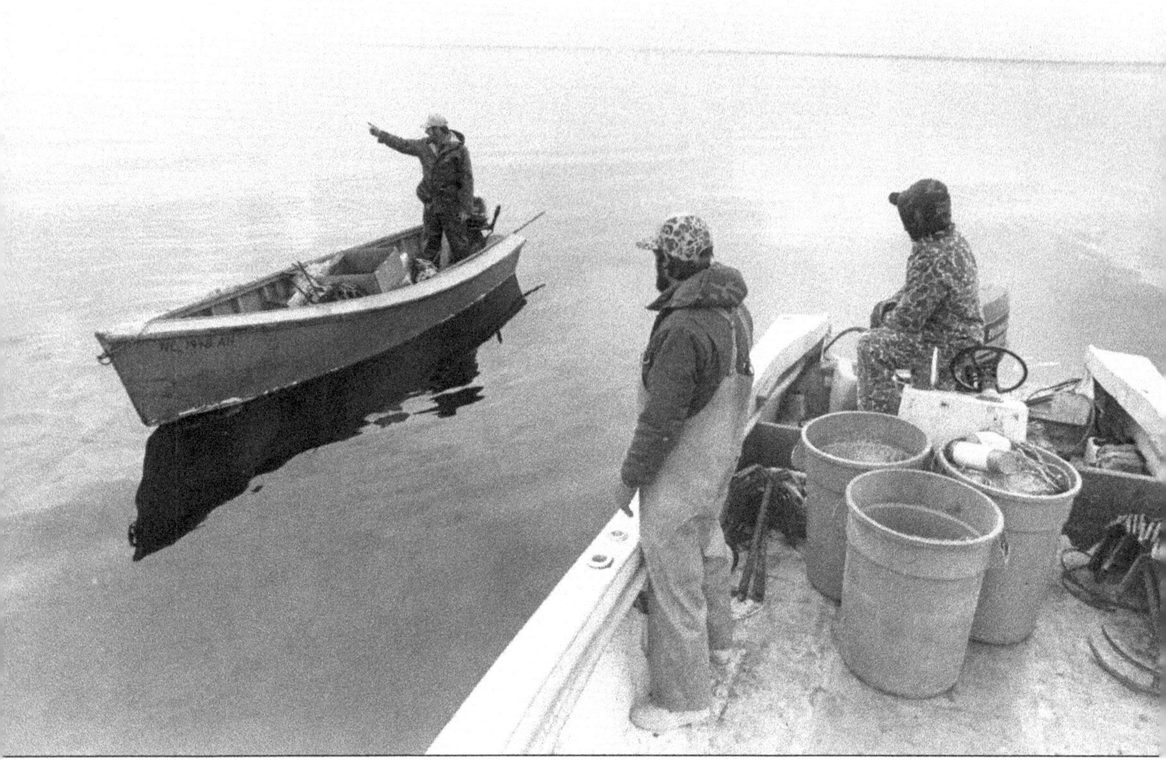

One of the first fisheries established on the Outer Banks was mullet fishing. As early as 1588, Thomas Harriot, who accompanied Sir Ralph Lane on his expedition to Roanoke Island, wrote that colonists were eating mullet. About 100 years later, salted mullets were sold or bartered for corn across the sound to settlers in Hyde County. Outer Bankers acquired a reputation for carefully cleaning, salting, and packing their catches in wooden kegs for far-flung destinations. No one knows why mullets tend to leap out of the water, but fishermen get excited when they see, or sometimes first hear, a school splashing in the water. Modern flat-bottomed boats have an elevated seat so the driver can spot schools more easily. The outboard motors are mounted in a well to keep the engine and prop away from lines and nets. On a slick-calm day, these fishermen conferred about the whereabouts of mullets. (Courtesy of Drew Wilson Collection, OBHC.)

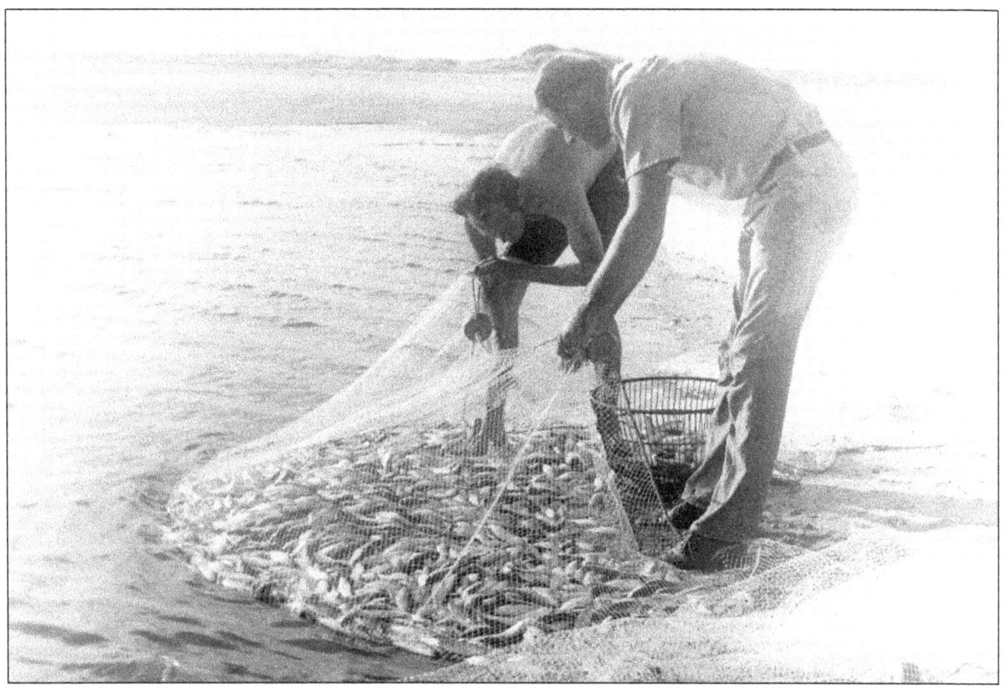

These corncob-size mullets were probably used for bait, but they are also delicious for human consumption. Mullets possess a gizzard that locals consider to be a delicacy. "Tobacco Bill" Austin, of Hatteras, kept cold fried mullet gizzards in his pants' pocket, offering them to friends and acquaintances; if his offer was rejected, he popped them in his mouth and ate them. (Courtesy of Aycock Brown Collection, OBHC.)

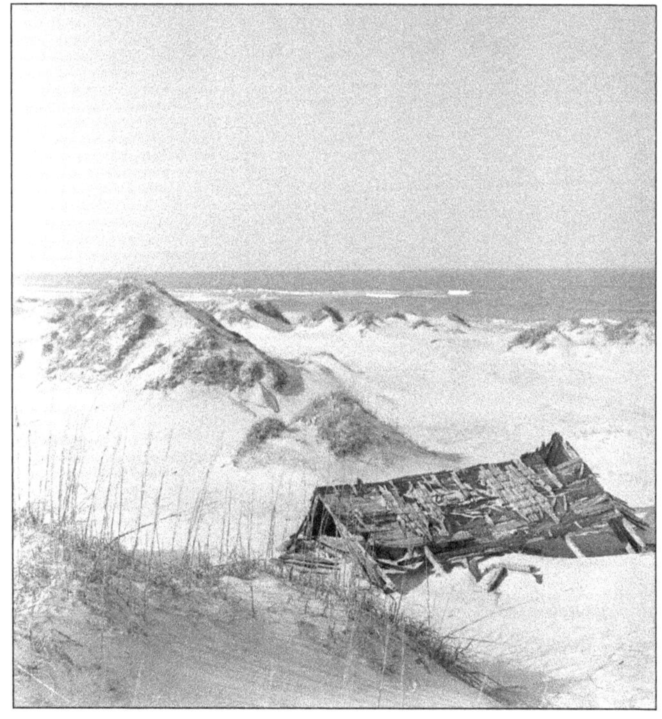

Photographer Aycock Brown captured a bit of forgotten history in the 1940s when he took this shot of a dilapidated fishermen's net house. Staying away from village homes for several days at a time to fish on the bare beach, they built shacks to use as camps for cooking and sleeping. Inside, they also stored their nets and other gear until the next set. (Courtesy of Aycock Brown Collection, OBHC.)

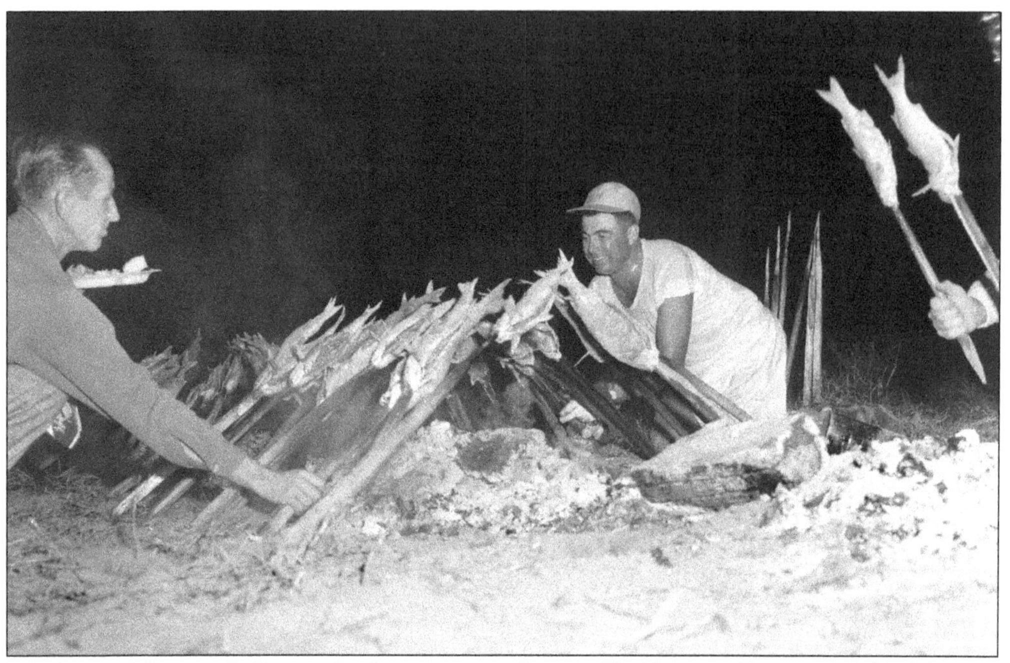

Residents of Currituck County had a unique tradition of holding mullet roasts. The fish were impaled from mouth to tail on thick green sticks that were stuck in the sand over hot coals from a wood fire. Each row of sticks leaned at a 45-degree angle, bracing and holding up another mirror-image row. (Courtesy of Aycock Brown Collection, OBHC.)

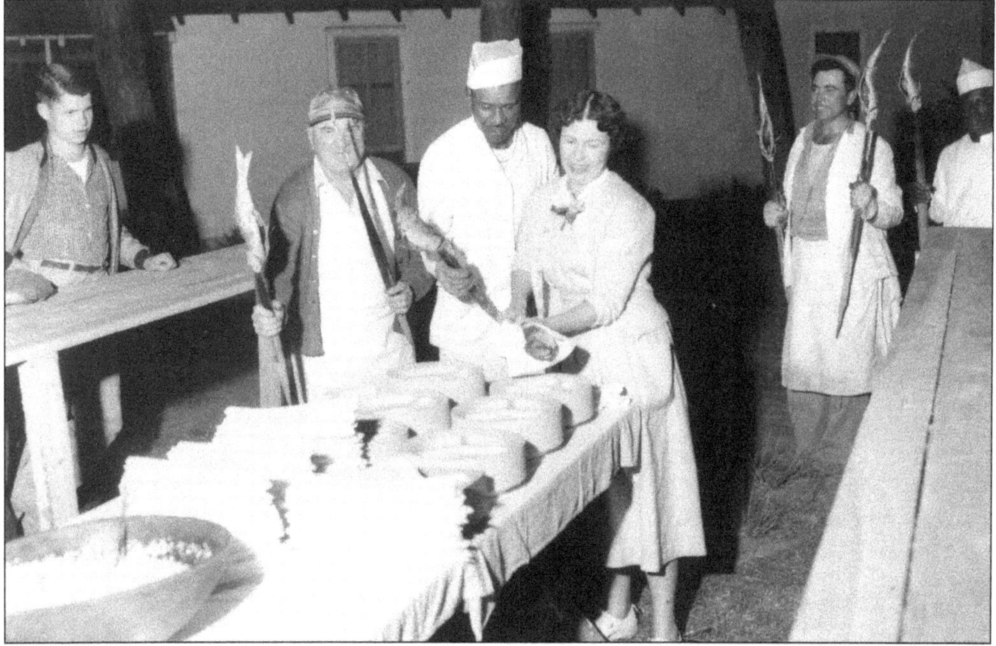

The Currituck downhome mullet roast was kicked up a notch when it became an affair for entertaining visiting dignitaries. In Point Harbor, men in suits and women in high heels enjoyed having their mullets roasted and handled by neatly dressed servers. (Courtesy of Aycock Brown Collection, OBHC.)

Author and historian David Stick wrote that the three quarters of a century between the Civil War and World War II was the great era of commercial fishing. Fishermen, like Ulysses Meekins, were able to catch and sell all that the market could bear. The long wharf, net house, and boats behind his Colington home bore testimony to a thriving family business. (Courtesy of Lish Meekins.)

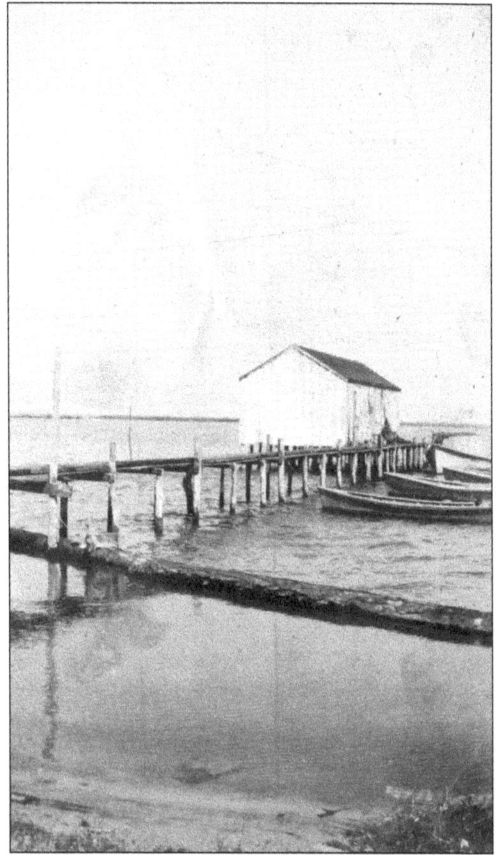

There is something addictive about catching jumping mullets like these newly caught mullets iced down in a basket at O'Neal's Sea Harvest in Wanchese. On the morning after his mother passed away, one mullet fisherman accepted a friend's condolences, and then said, "Yes, it's sad. Hey, you should have seen that mullet catch I had last night!" (Courtesy of RWG.)

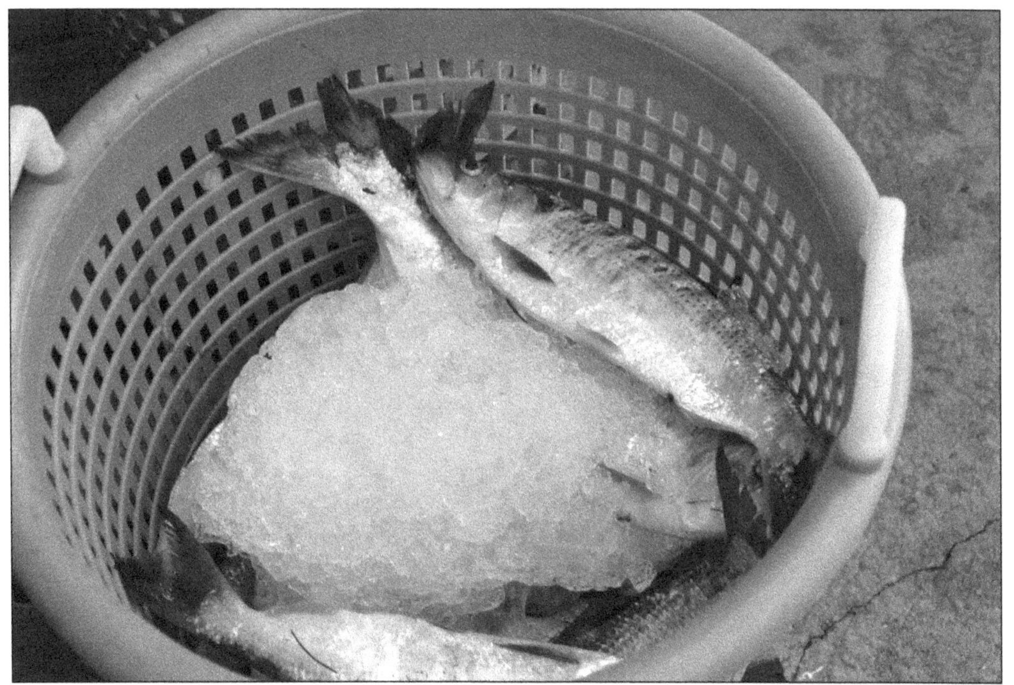

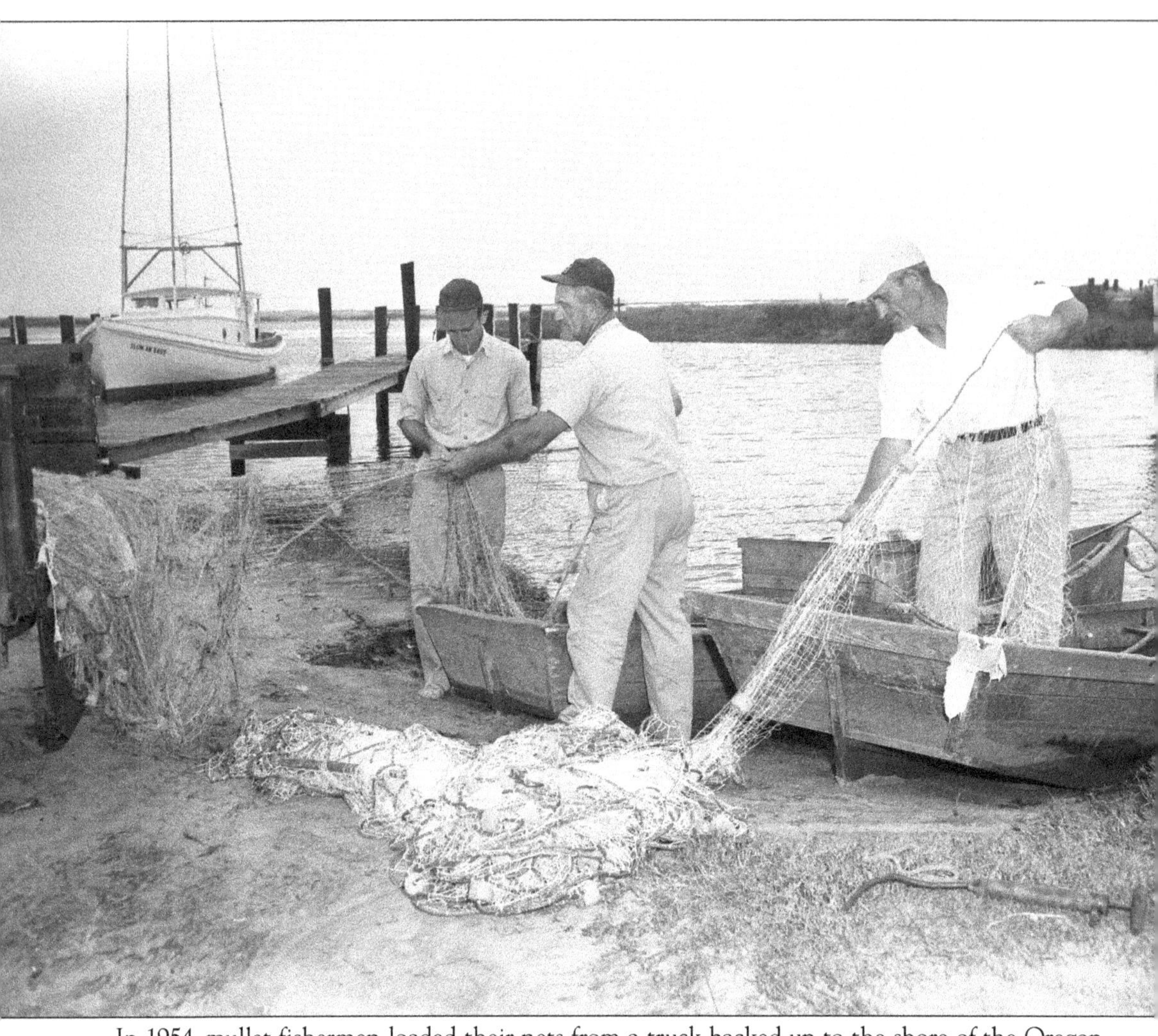

In 1954, mullet fishermen loaded their nets from a truck backed up to the shore of the Oregon Inlet boat basin, a place where today only sportfishing charter boats dock. Barefoot and dressed in work khakis, the fishermen's boats were as simple as their style of dress. Since mullet fishing usually takes place in shallow water, a very basic boat can be used. These boats probably had an inboard automobile motor that was converted for marine use. More often than not, there was no transmission, so timing was everything when approaching a dock. When working in the sound, the fishermen drove a net staff in the sandy bottom and ran the net out of the boat to encircle the mullet that flapped and flipped about on a shoal. (Courtesy of Aycock Brown Collection; OBHC.)

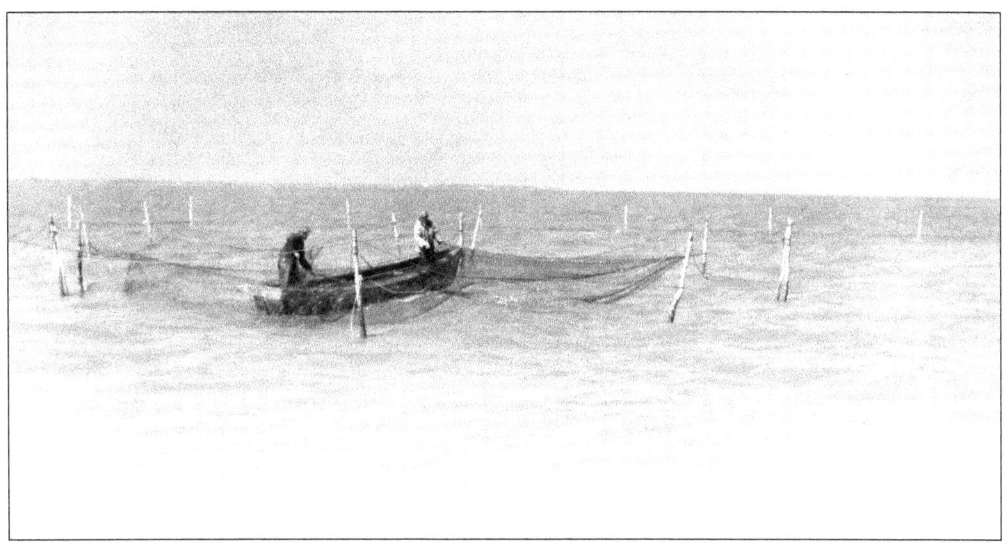

Working shad fishermen were captured on film by a visiting photographer to Wanchese in the 1930s. The men were inside of the pound net bunching up the shad before scooping them out with an iron-ringed dip net. Just like the weirs of the Algonquin Indians, pound nets have a tunnel that funnels the fish into a larger enclosure that they cannot easily escape from. (Courtesy of SANC, photograph by Charles Farrell.)

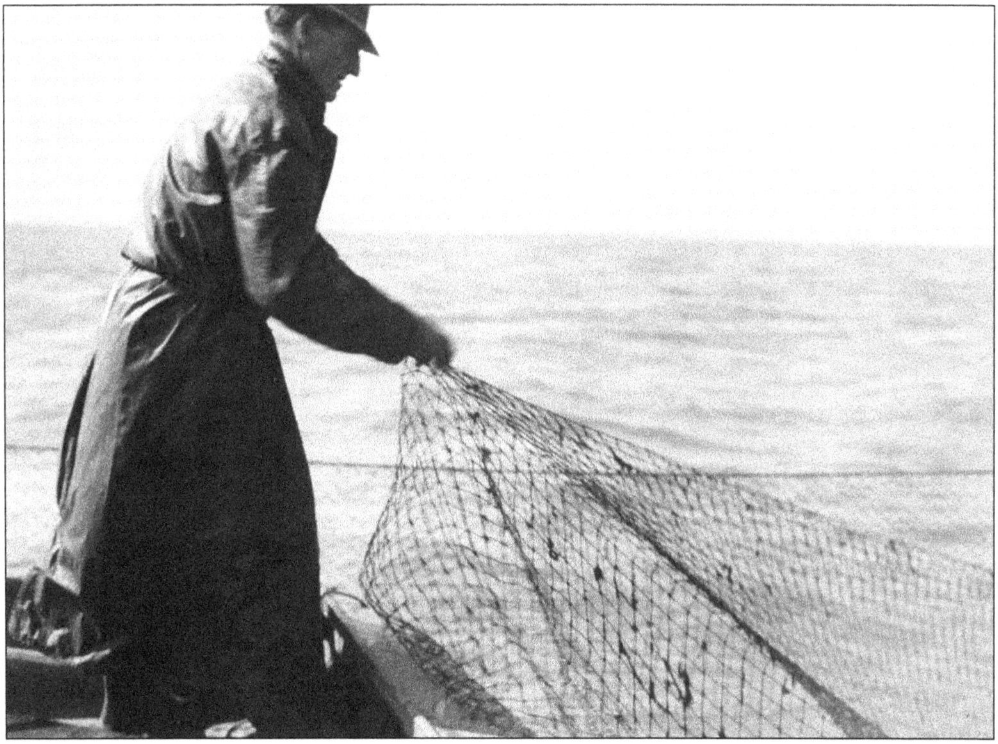

Shad fishing was an economic boon to the area, with the apex in the 1880s and 1890s. A year's wages could be made during one season of backbreaking work. Some crews stayed in crude camps in the Roanoke marshes for quicker access to fishing grounds each day. Over 100 camps once stood on land that has now washed away. (Courtesy of SANC, photograph by Charles Farrell.)

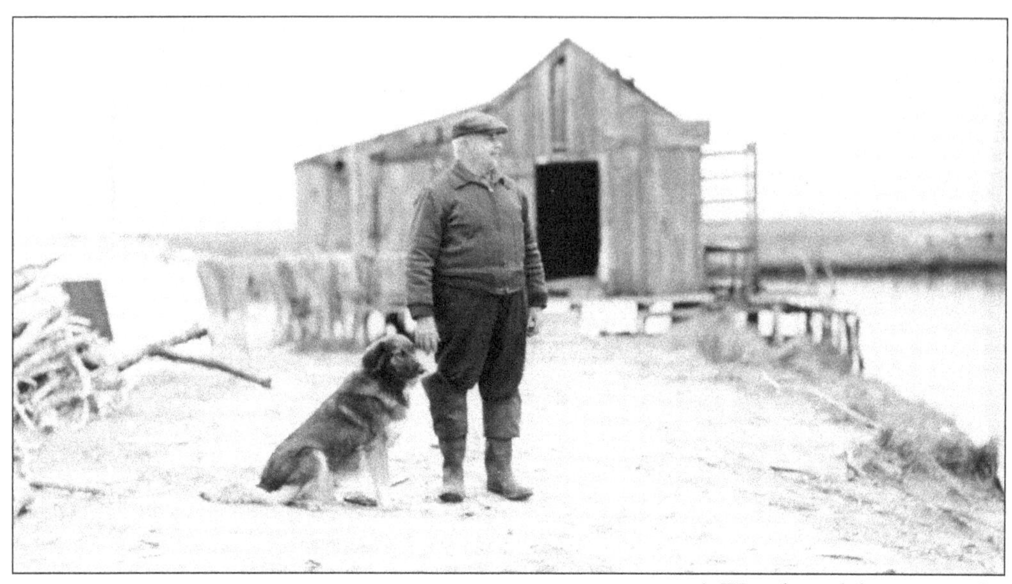

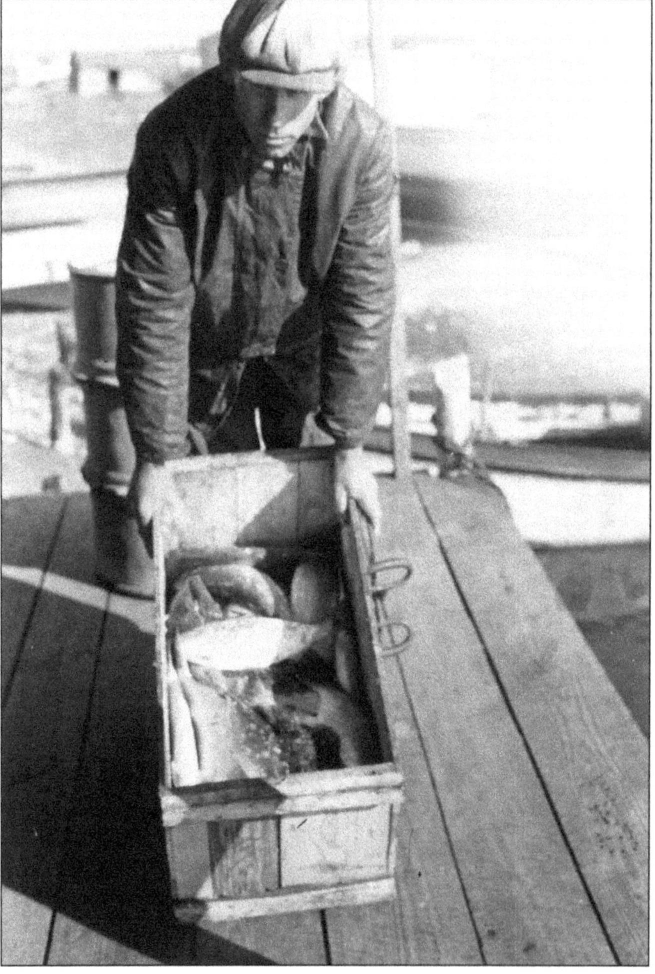

A Wanchese fisherman and his best friend stood near a stack of wood that might become pound net stakes. Wayland Baum, who once fished 10 pound nets each season, said, "I could stick stakes in my sleep almost I done so much of it." He had to hang lighted beacon lanterns to keep boats from going through his nets that nearly blocked off the whole sound. (Courtesy of SANC, photograph by Charles Farrell.)

With ice tongs dangling from one hand, this fisherman shows off his catch of shad. One of the most richly flavored fish, shad is also one of the boniest. Spawning females entered the inlet and came up the sounds full of roe, the internal egg mass, which was the moneymaker for watermen. A shad roe cannery was built midway between Wanchese and Manteo at Skyco. (Courtesy of SANC, photograph by Charles Farrell.)

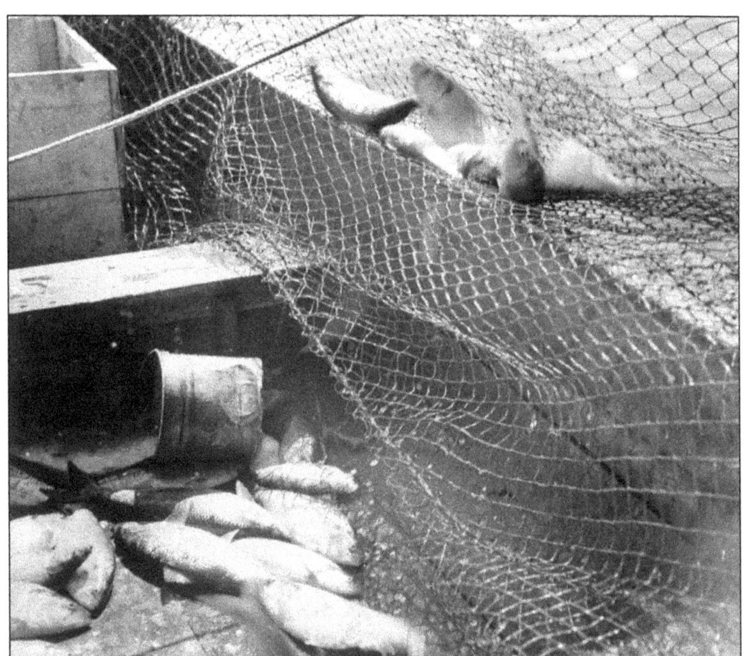

The lowly shad was the catalyst for a remarkable amount of social and economic change on the Outer Banks. As a result of historic seasonal runs, by 1870, four shad fish houses emerged on the southern part of Roanoke Island. Wanchese became a bustling town with multiple stores staying open until 11:00 at night. The lean years of the Civil War gave way to prosperity and hope. (Courtesy of SANC, photograph by Charles Farrell.)

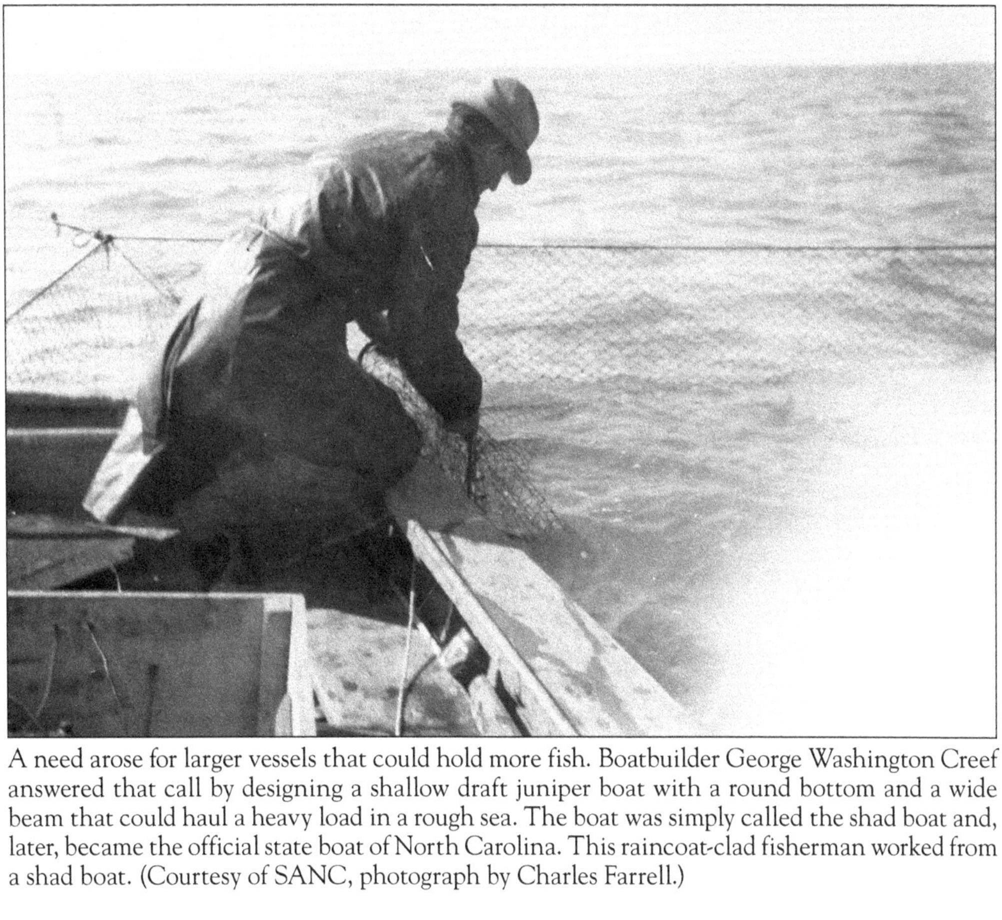

A need arose for larger vessels that could hold more fish. Boatbuilder George Washington Creef answered that call by designing a shallow draft juniper boat with a round bottom and a wide beam that could haul a heavy load in a rough sea. The boat was simply called the shad boat and, later, became the official state boat of North Carolina. This raincoat-clad fisherman worked from a shad boat. (Courtesy of SANC, photograph by Charles Farrell.)

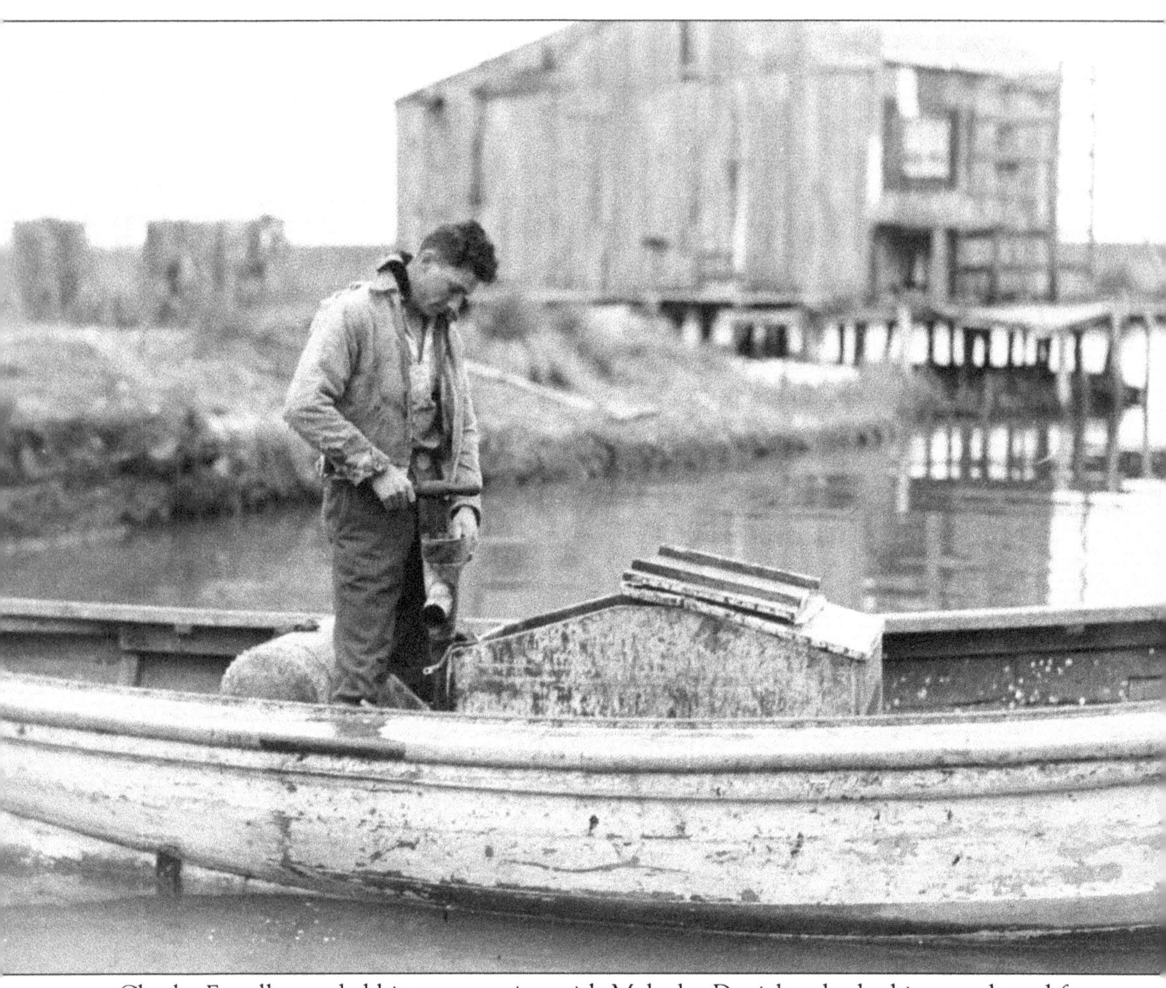

Charles Farrell recorded his conversation with Malcolm Daniels, who had just graduated from high school. Pumping out his father's shad boat by hand, he told the journalist that he did not want to fish for a living and asked him, "Do you know where I could find a job?" As it turned out, Daniels did not have to fish his whole life and became a fish buyer and the founder of Wanchese Fish Company in 1936. Using his visionary mindset, he transformed used junk into fishing trawlers, fish house equipment, and freight trucks, finding a way to put Wanchese on the map as a hub of seafood transportation. He and his wife, Maude, had 15 children who all worked hard in the family business and who later all benefitted from their parents' tenacity and prayers when the company went global and sold to a Canadian firm for over $65 million in 2015. (Courtesy of SANC, photograph by Charles Farrell.)

Black and white crewmembers worked together side by side with little racial tension. All fishermen, like William Pell, were judged by how hard they could work. The booming shad fishery was also responsible for the first ice plant being built on Roanoke Island in Skyco because ice could not be shipped in fast enough to keep the massive catches from spoiling. (Courtesy of SANC, photograph by Charles Farrell.)

Seagulls showed great excitement over 22 pogy, or menhaden, boats that took over tiny Silver Lake Harbor in Ocracoke in November 1957. Large trawlers from other ports, like Beaufort, Wilmington, or Southport, North Carolina, came to sweep up the oily fish and carry them back to fertilizer processing facilities. (Courtesy of NPS, CHNS.)

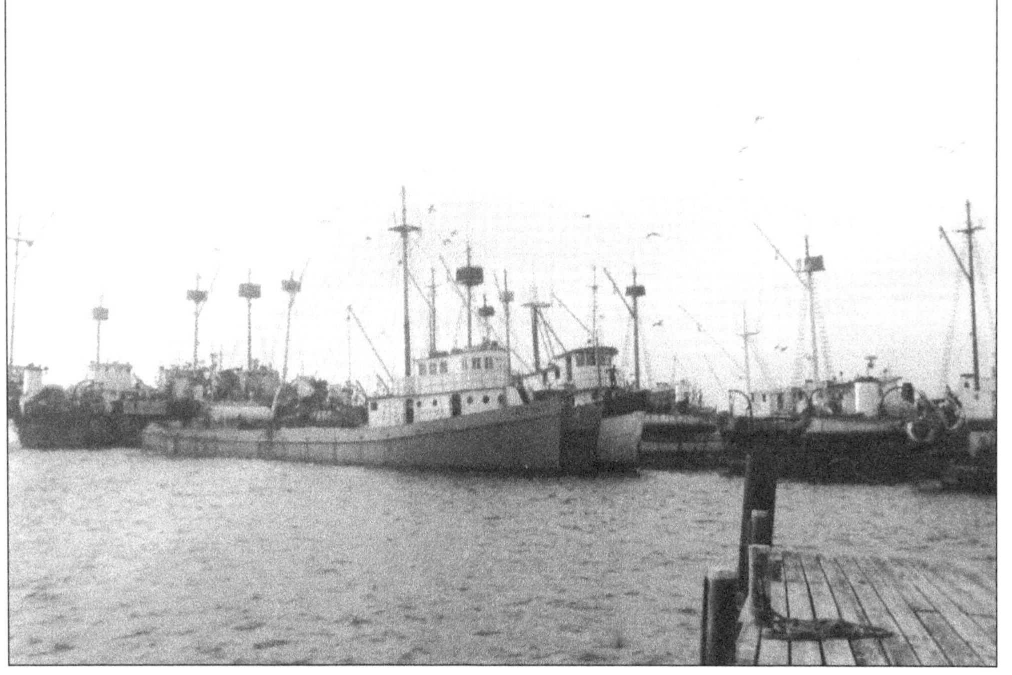

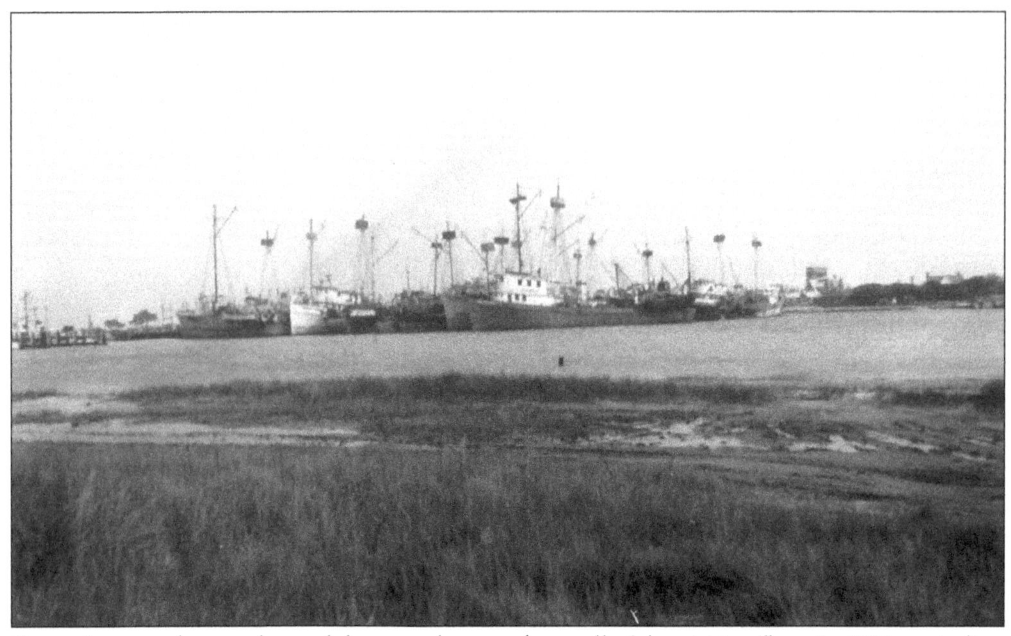

Ocracokers may have welcomed the crews but not the smell of the visiting fleet. In 1866, capitalists prospecting the area for potential menhaden processing plants reported that they were driven away by natives of Roanoke Sound. Nonetheless, in the late 1800s, plants were established at Portsmouth Island, Oregon Inlet, and Roanoke Island that were abandoned at a loss after two or three years. (Courtesy of NPS, CHNS.)

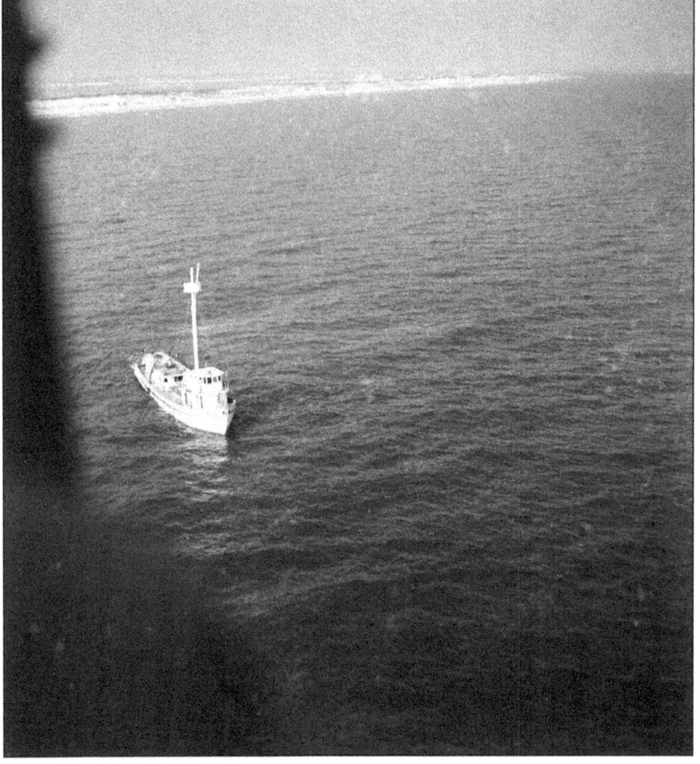

This aerial shot of a menhaden trawler might have been taken from an accompanying airplane used to spot large schools. Part of the word menhaden is derived from an Algonquin word meaning "he fertilizes," and it is generally believed that Tisquantum, also known as Squanto, advised the pilgrims to plant menhaden with their crops. (Courtesy of Aycock Brown Collection, OBHC.)

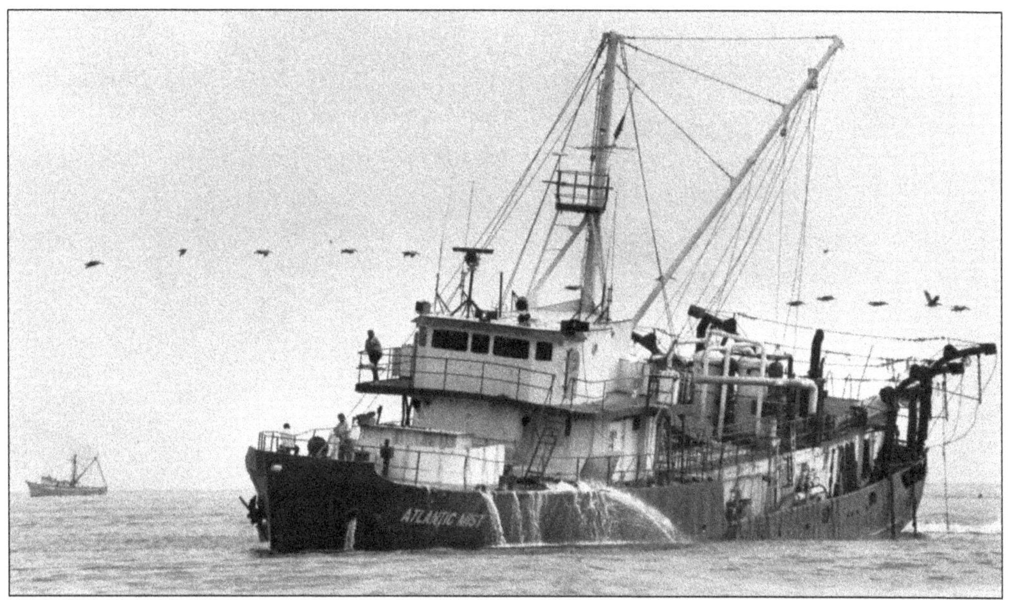

The captain and crew of the 177-foot-long *Atlantic Mist* gained firsthand knowledge of the Outer Banks' reputation as the Graveyard of the Atlantic when their vessel struck a submerged object that tore three gashes in the hull on November 27, 1990. The menhaden boat owned by Ampro Fisheries of Reedsville, Virginia, listed about 200 yards off the beach at Rodanthe. (Courtesy of Drew Wilson Collection, OBHC.)

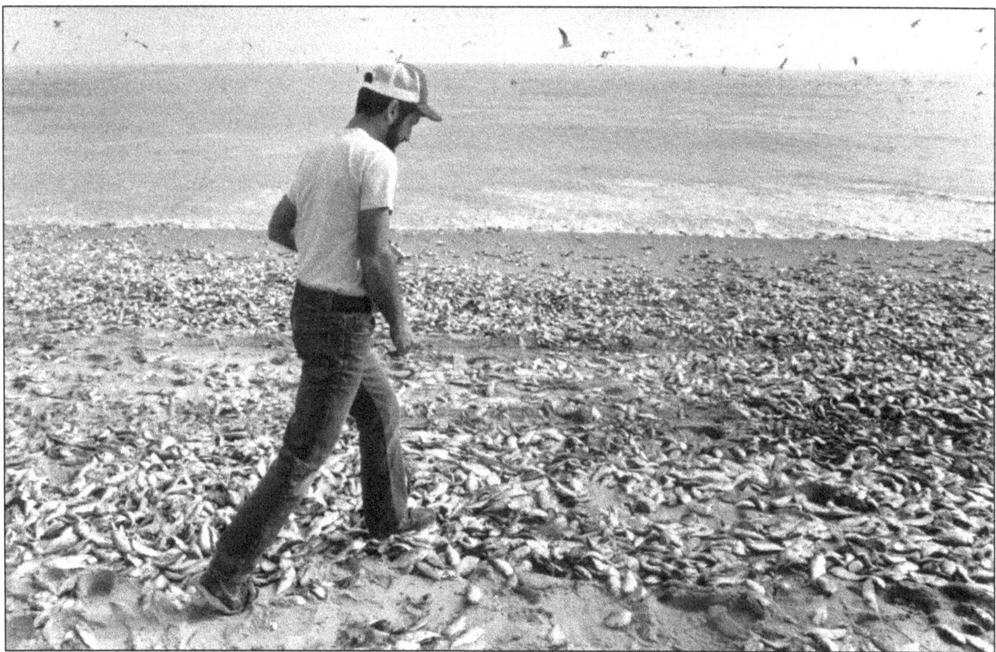

The *Atlantic Mist* was forced to dump thousands of pounds of menhaden and other fish after it started taking on water. Larry Lamborne of Crofton, Maryland, walked through the decaying loss. Menhaden, also called fatback, are used for other purposes besides fertilizer, including the manufacture of meal for animal food and the extraction of oil to be used in paint. (Courtesy of Drew Wilson Collection, OBHC.)

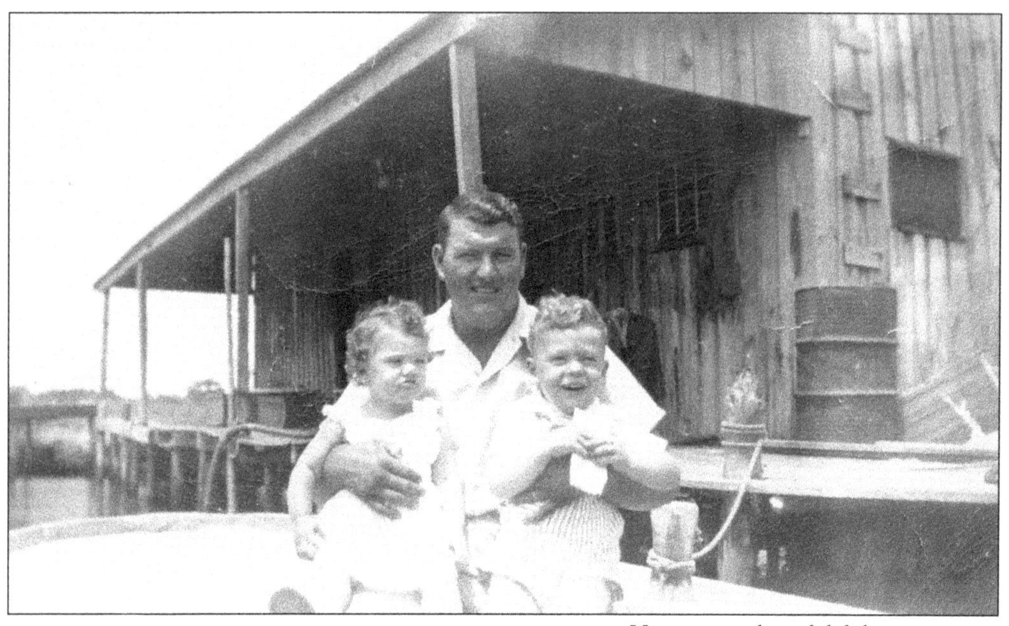

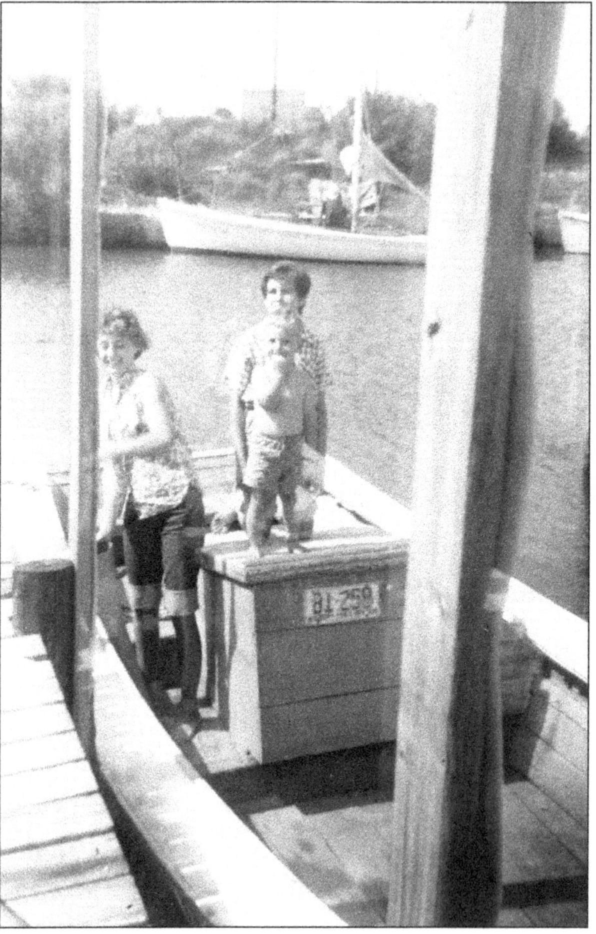

Herring, a plentiful fish in the same family of shad and menhaden, was a cheap source of food high in protein and calories. Long net fisherman Macon Meekins, pictured with his children Myrtle and Charles, teamed up with his brother Ralph to bring in herring. The men provided well for the 12 children between the two of them. (Courtesy of Joyce Meekins.)

Macon Meekins's children, from left to right, Yvonne, Elisha, and Minta clambered aboard their father's boat docked in Wanchese in 1964. Later, two of Macon's sons worked with him until his death in 1986. The family no longer fishes for a living; however, Minta managed Oregon Inlet Fishing Center for over 40 years. (Courtesy of Joyce Meekins.)

At left in this image, Lish Meekins takes a Dr Pepper break before using his young muscles to bring the net into the boat. Elmer Sawyer (in a striped shirt) and Macon Meekins get a glimpse of what might have been caught that day in the mid-1970s. Long net fishermen accepted the fact that they could pull 2,000 yards of net and either get a gracious plenty or a measly bucket full. (Courtesy of Joyce Meekins, photograph by Reverend Luther Wesley.)

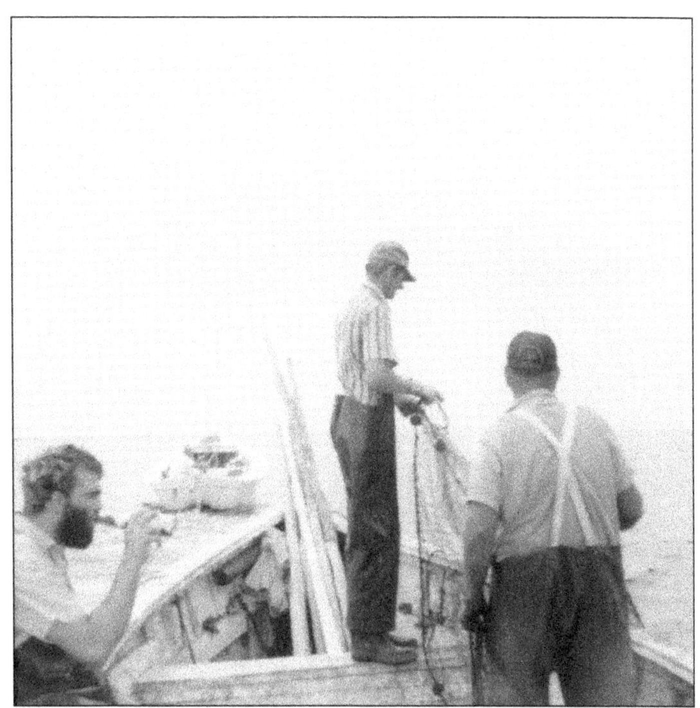

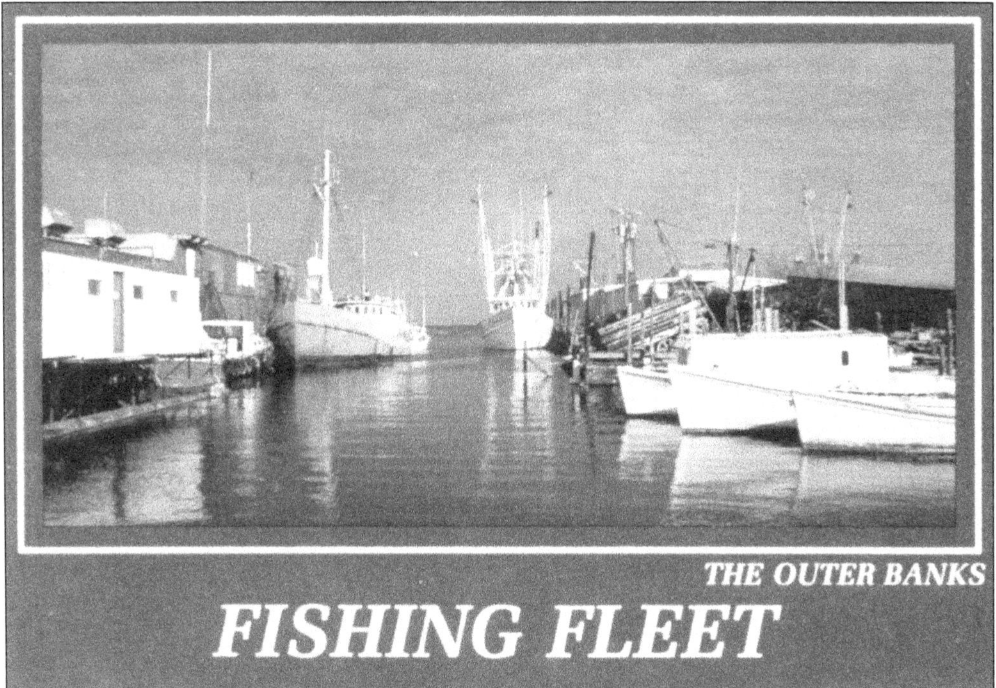

A postcard captured the Meekins family's long netting rig (on right) at rest near the public ramp in Wanchese harbor. The cabined vessel, a converted shad boat, provided a place to cook and sleep when the men were fishing down in Hatteras. The run boat helped set the net and, ideally, carried a big haul to market. The skiff held the yards of net and sticks to keep the net from rolling up. (Courtesy of Joyce Meekins.)

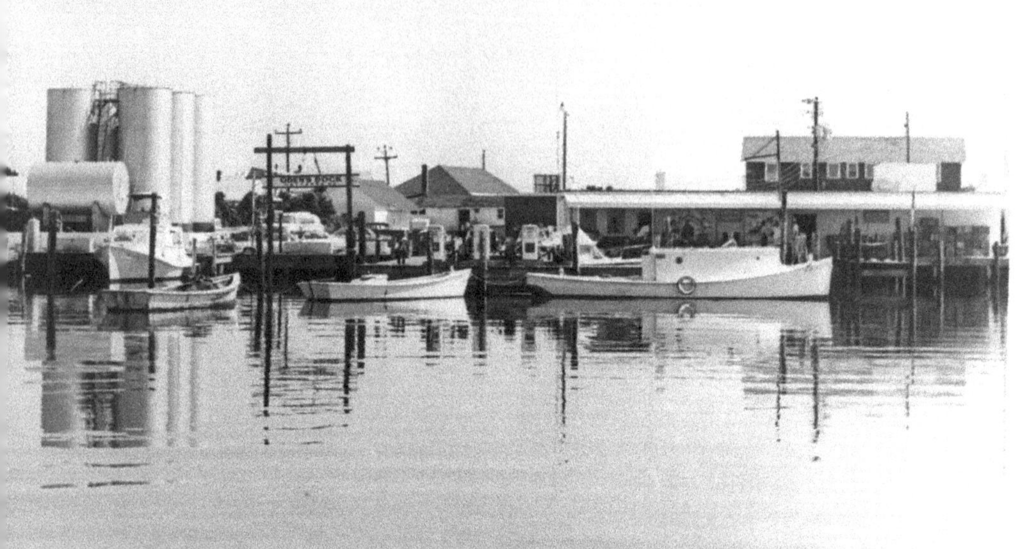

Inshore fishermen sold their catches at Oden's Dock in Hatteras Village, back when huge gas tanks dominated the waterfront. These days, Oden's accommodates commercial fishermen by selling fuel and renting dock space, but there is no fish house on the property. Tourist-oriented charter boats and head boats take precedence. According to a North Carolina Sea Grant inventory, the presence of seafood wholesale facilities is rapidly declining due to decreasing profits, increasing operating costs, labor shortages, stricter fishing regulations, declining water quality, and scarcer fish stocks. Certainly in Dare County, developmental pressures factor in as a once sparsely populated area swells with visitors and year-round residents willing to pay a premium for seaside property. (Courtesy of Joyce Meekins, photograph by Aycock Brown.)

Three

PRIZED CATCHES

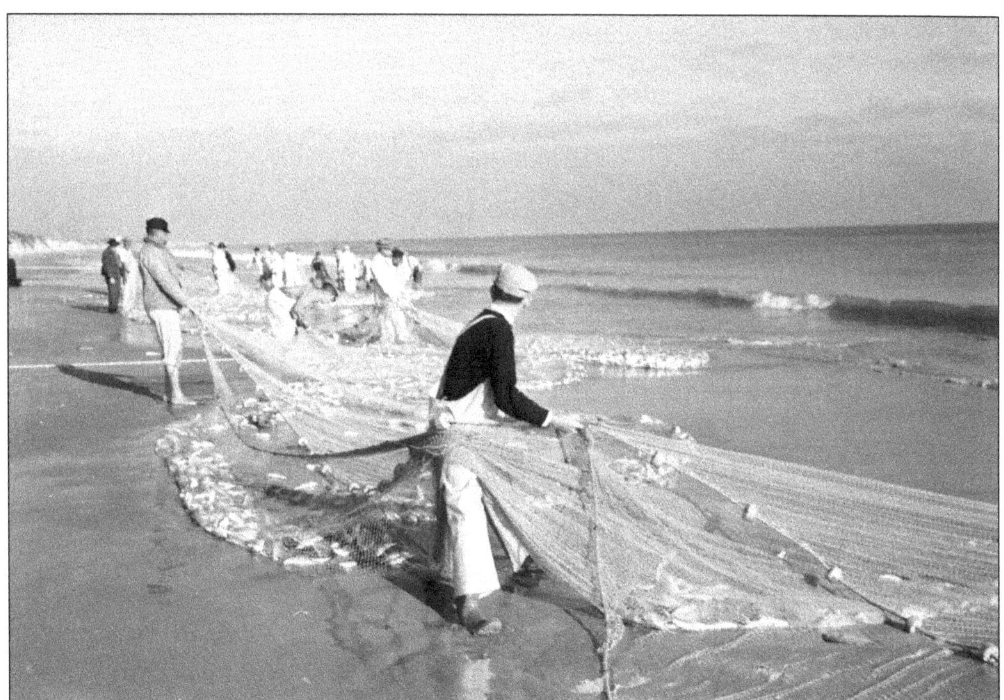

It takes a village to pull a net in by hand, especially a heavy cotton net full of croakers in the 1950s. A sharp-bowed, flat-bottomed boat called a dory was used to make the set. The shore end of the net had a long, attached rope that extended through the breakers and was anchored near the dunes. The bunt, or bagging end, of the net was "played out" of the stern of the dory as it went to sea. (Courtesy of Aycock Brown Collection.)

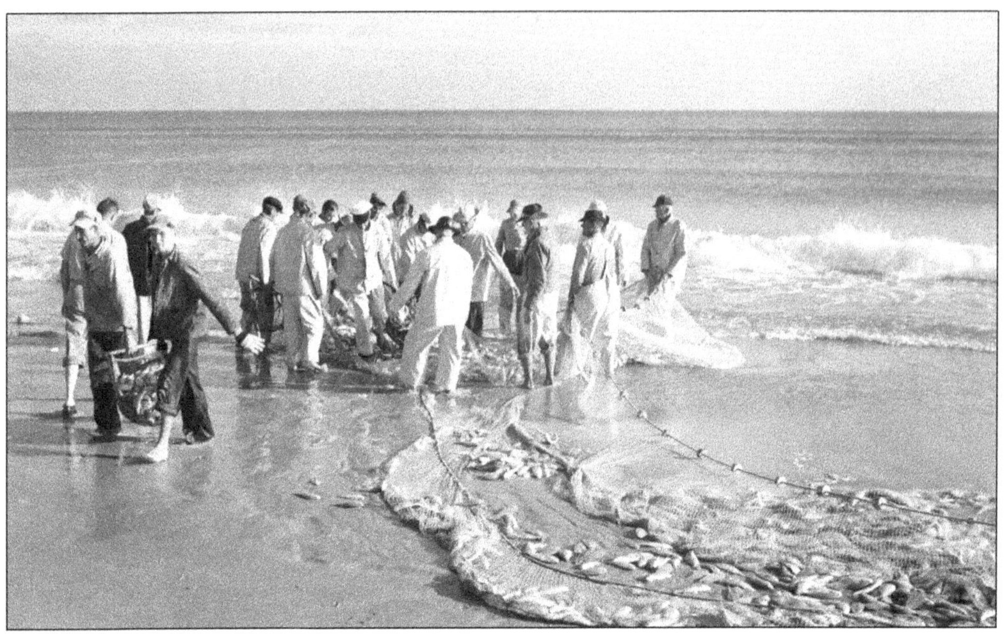

The dory crew positioned the beach seine in a half-moon and secured it with a special offshore anchor they would later trip when it was time to haul in the net. When they came to shore, the men used a third anchor to fasten the wing end of the net to the beach. The men in the photograph may have been reenacting the old practice of hand hauling, since vehicles were used to pull nets in the 1950s. (Courtesy of Aycock Brown Collection.)

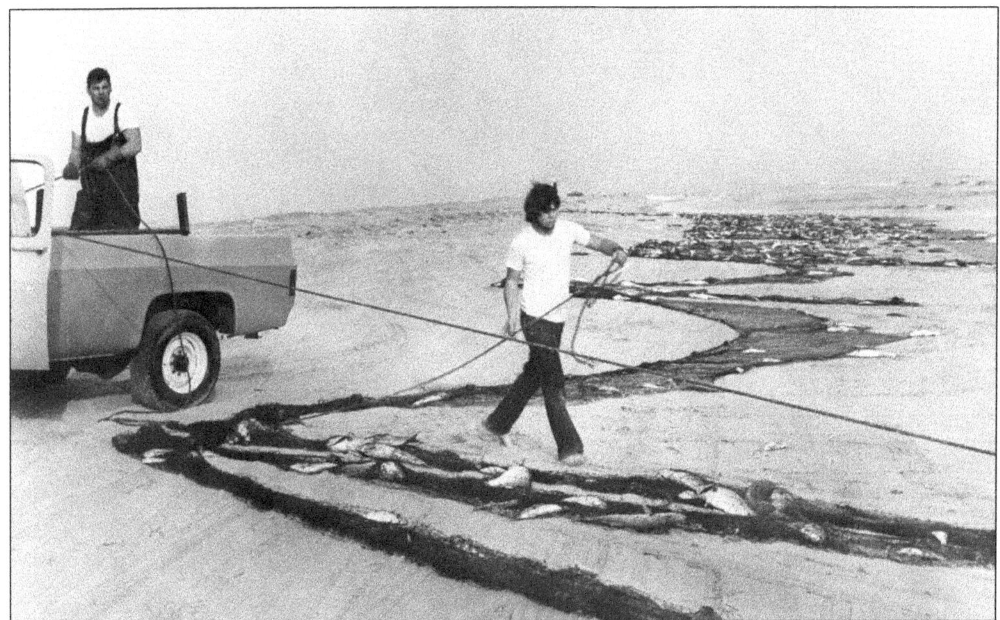

Just two men using a four-wheel drive truck to haul in a net was a more efficient utilization of manpower and time. "Tie ropes" were knotted to the net and attached to the truck and repeatedly tied and released to pull in the catch a little at a time. Beach nets were usually set in the evening and pulled in the morning. Another crew with dories and a pickup truck can be seen north of this one. (Courtesy of Aycock Brown Collection.)

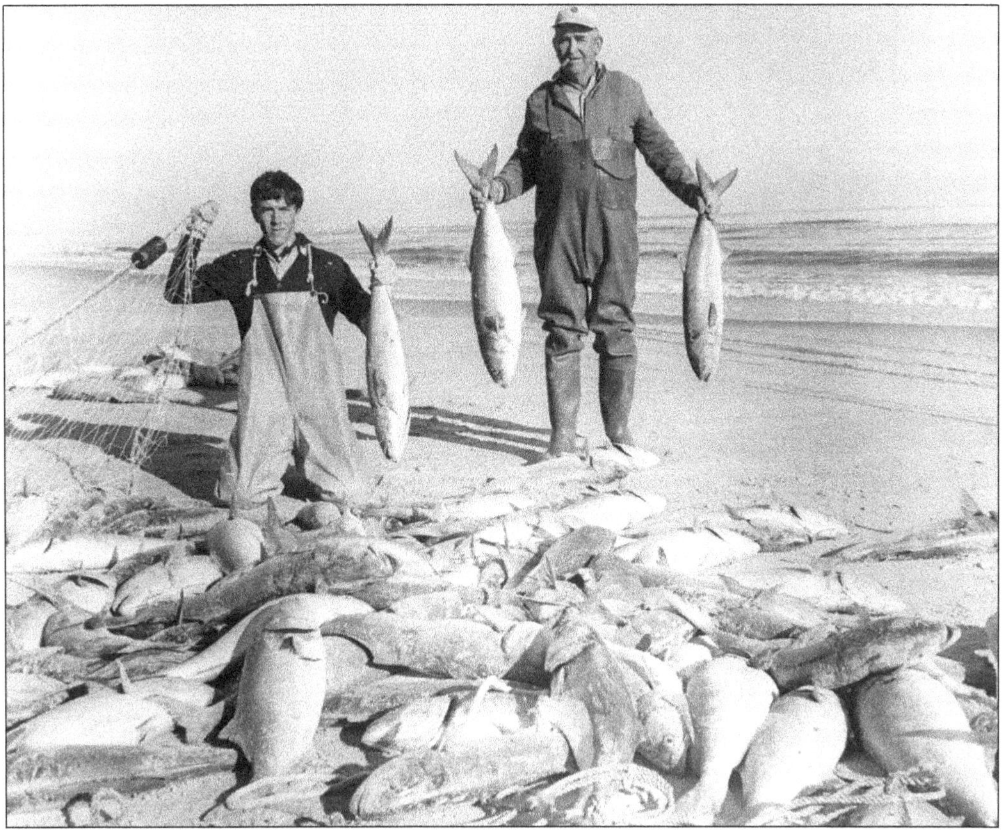

A bounteous catch of big bluefish was an excellent start to the day. An experienced beach fisherman would set his net in the evening and place an empty water glass on his windowsill when going to bed. If the wind shifted in the night, the glass would make a clatter as it blew off the sill, and the fisherman knew he must get up and pull in his net before it became a twisted mess. (Courtesy of Aycock Brown Collection.)

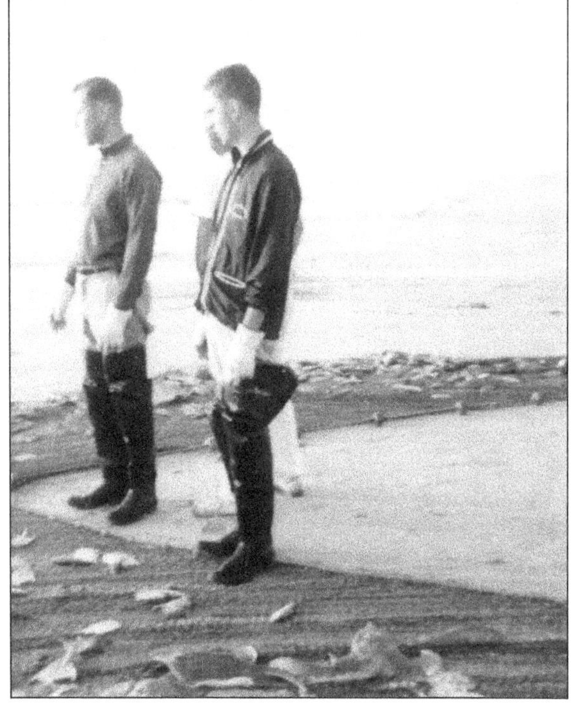

Cousins Jeffrey Midgett (left) and coauthor Wayne Gray look none too thrilled at the morning's work ahead of them. The young men beach fished with Jeffrey's father, Jethro Midgett Jr., at daybreak and then cleaned fish and worked in the family retail market for the remainder of the day. This catch included skates, seen in the foreground, which would be thrown back. (Courtesy of Jeffrey G. Midgett.)

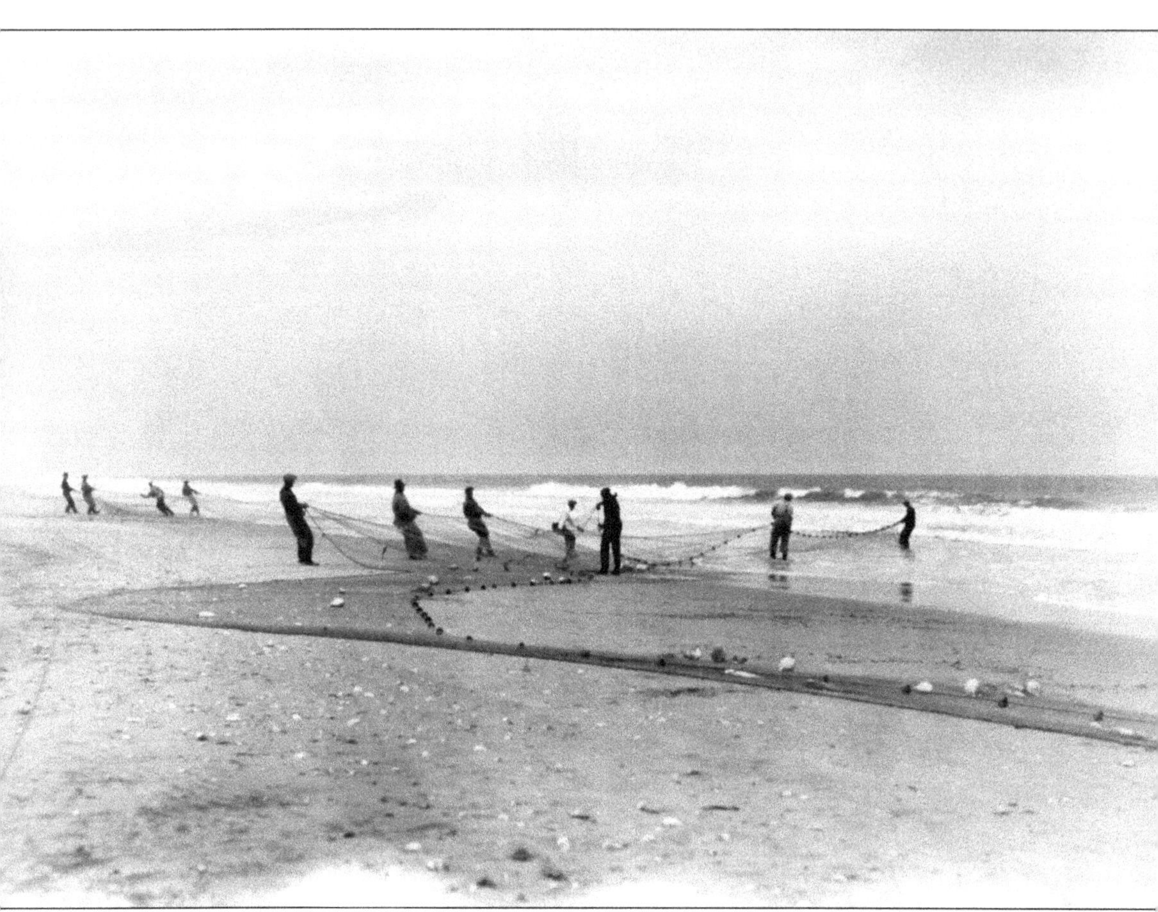

Although the bunt end had not yet been brought up, rewards this day did not look very promising. Puffer fish, or blow toads, spaced few and far between, would not be enough. Each crewmember got a share of the profits, with the captain and the boat getting a larger percentage. Beach seine net hauling was one of the earliest methods of fishing on the Outer Banks. The standard sized net was 300-yards long, 14 feet deep, with usually a three-inch mesh, according to what species was running. A dory was rowed out through the breakers and back to shore. As time progressed, fishermen employed inboard and outboard motors, but oars always stayed on the dories for almost certain engine failures. Hand hauling gave way to better pulling practices provided by surplus Army jeeps and four-wheel drive pickups. (Courtesy of Jeffrey W. Midgett.)

Jethro Midgett Jr. untangled a small dogfish from his sturgeon net in 1960. Four generations of the Midgett family fished in front of the Nags Head Beach Cottage Row Historic District before it had designated status. Mutual respect and cooperation ruled, as the longtime summer residents would help pick the fish from the net and then take some home for their breakfast. (Courtesy of Jeffrey G. Midgett.)

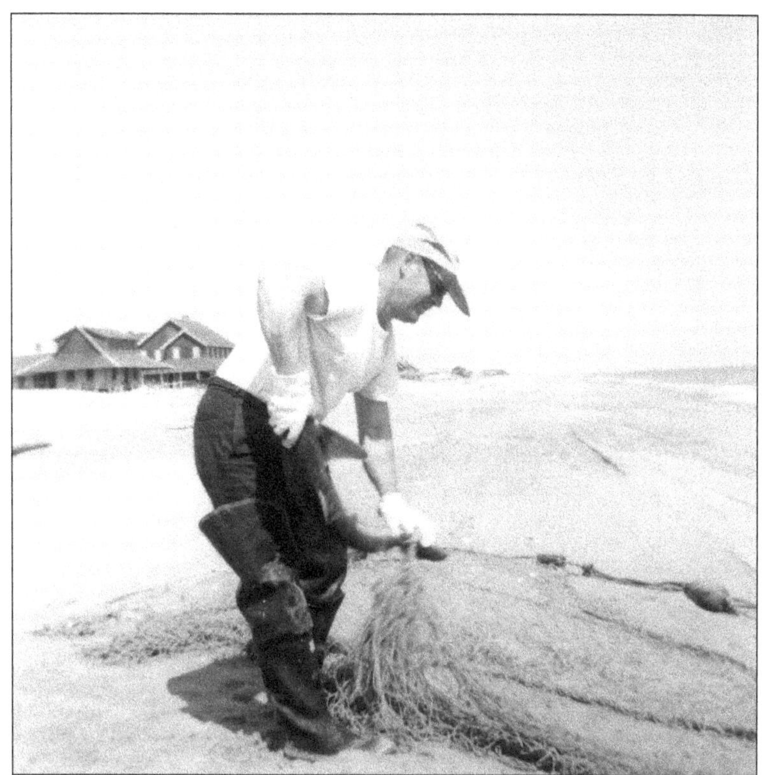

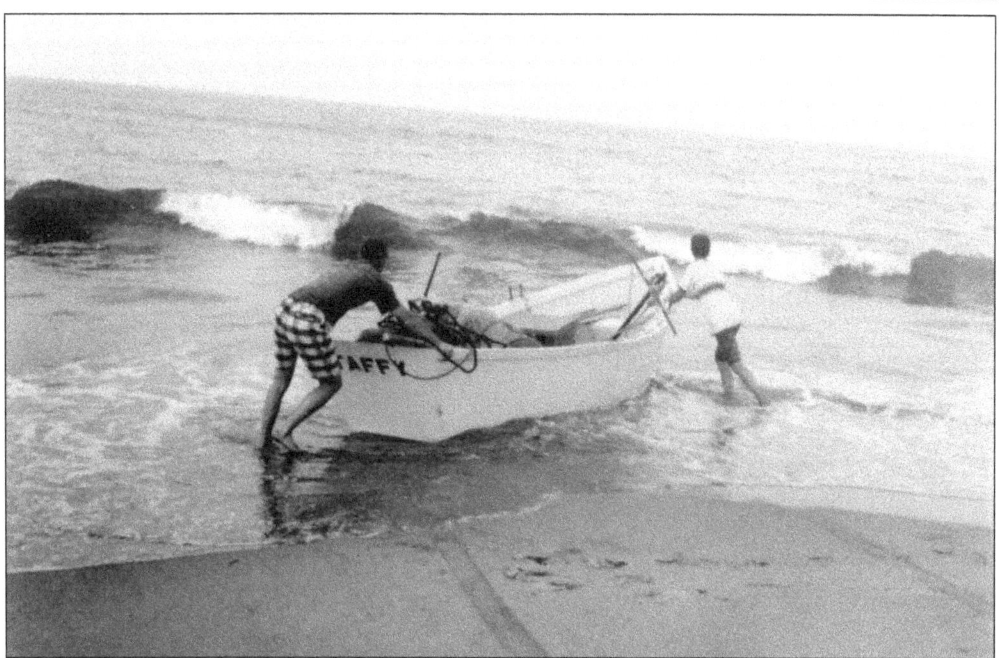

In 1966, tire tracks backed up to the wash show that the 16-foot dory, *Taffy*, had been off-loaded close to the water. The young man in the back lined up the boat, with the motor probably already running. The shore anchor to secure the net will be set, and the boys will jump in the dory and head to sea, gunning the motor as they hit the breakers. (Courtesy of Jeffrey G. Midgett.)

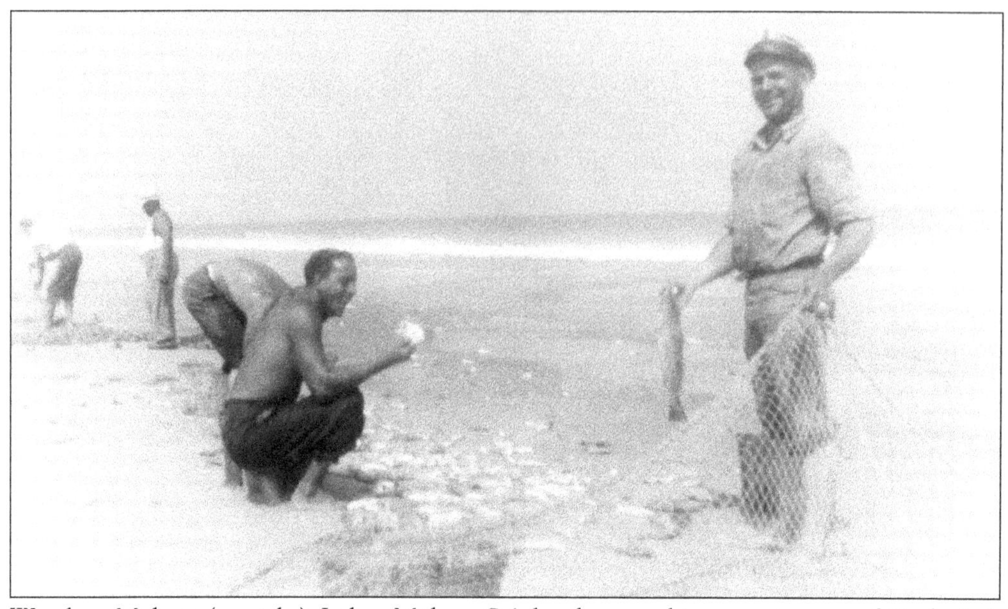

Woodson Midgett (on right), Jethro Midgett Sr.'s brother, ran his operation out of Duck. Jack Wise (crouching) seemed to laugh at the size of the fish they scored. Working in a more remote area, Woodson did not face the problems Jethro did. As early as the 1960s, beach fishermen had to contend with people aligned with recreational fishing cutting lines and setting the net adrift. (Courtesy of Jeffrey G. Midgett.)

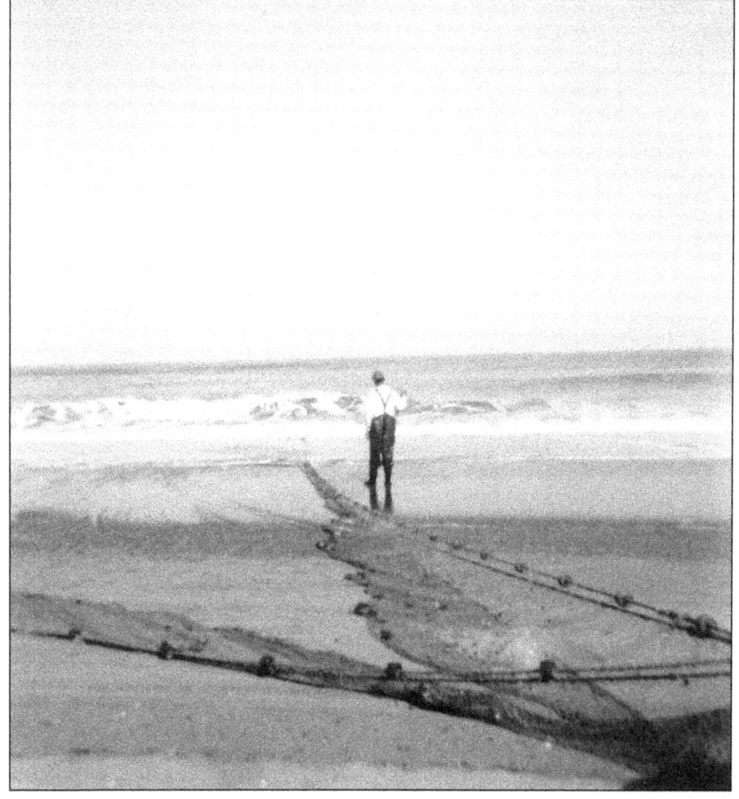

Charles Johnson waits for his work to begin on a calm October 1957 day. Clarence Brickle, a neighboring beach fisherman, suspected a cottage owner was sabotaging his gear. He waited, hidden, for several nights until the woman came along in her pink jeep. As she bent to cut the net line, Brickle frapped her on the rear end with a cedar shingle with all his might. It stopped the cuttings for a while. (Courtesy of Jeffrey G. Midgett.)

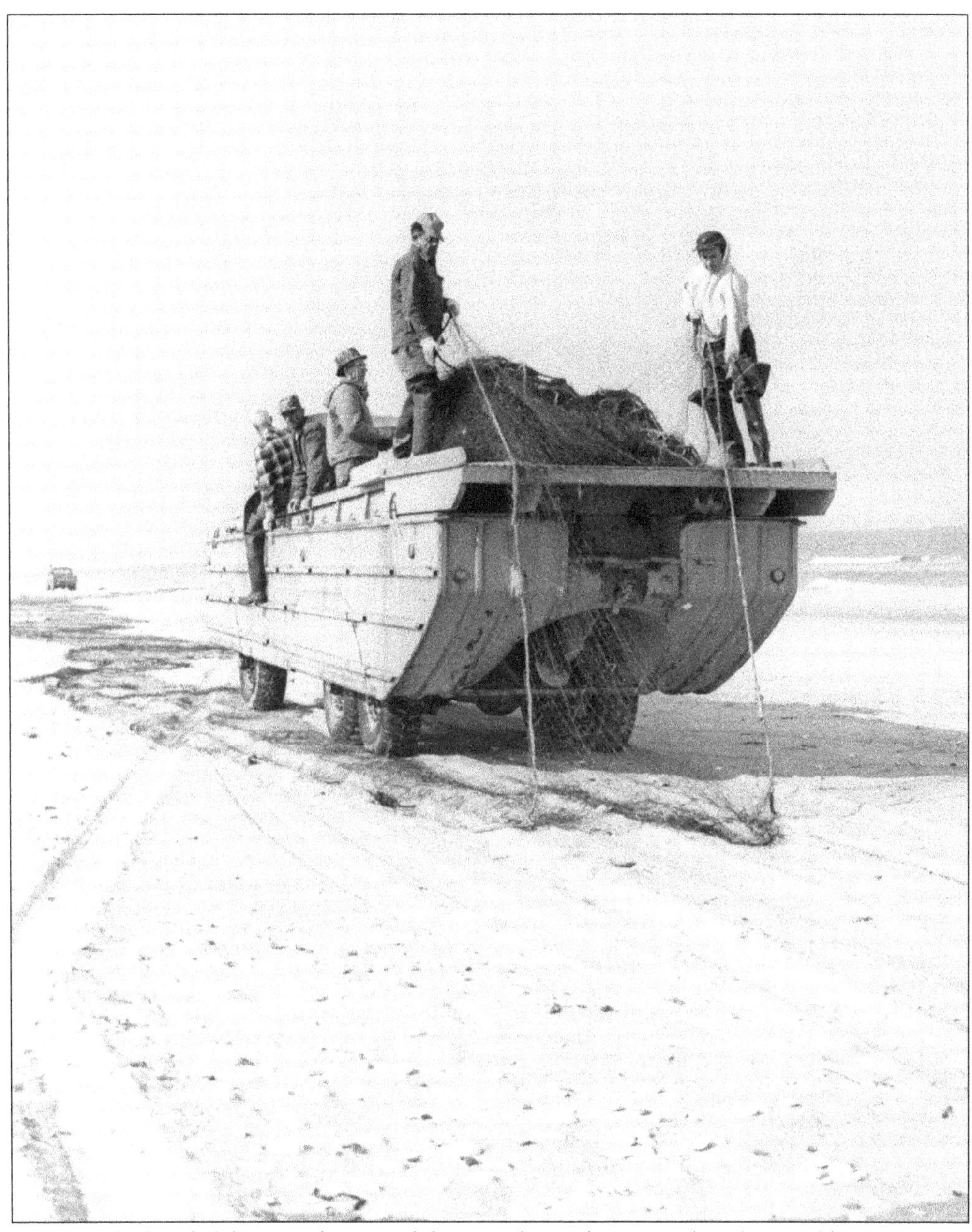

Outer Banks' beach fishermen discovered the joys of Army/Navy surplus after World War II. Many brought home treasures from the Norfolk, Virginia, area. The DUKW, pronounced "duck," seemed like the ideal vehicle for beach fishing. The name DUKW is not an acronym, but a manufacturer's code. D stood for 1942, U for utility, K for all-wheel drive, and W for two powered rear axles. The amphibious landing craft could set a net when the ocean was too rough for a lightweight dory, and some of the crew could operate from inside and be dry and relatively warm on a frigid day. Also, the novelty of the machines attracted a lot of attention from tourists. There was only one problem with this love match: the DUKW was out of commission for long stretches of time. In the end, the constant tinkering on the motors and the endless coaxing of them to run became too much for the fishermen. (Courtesy of Aycock Brown Collection, OBHC.)

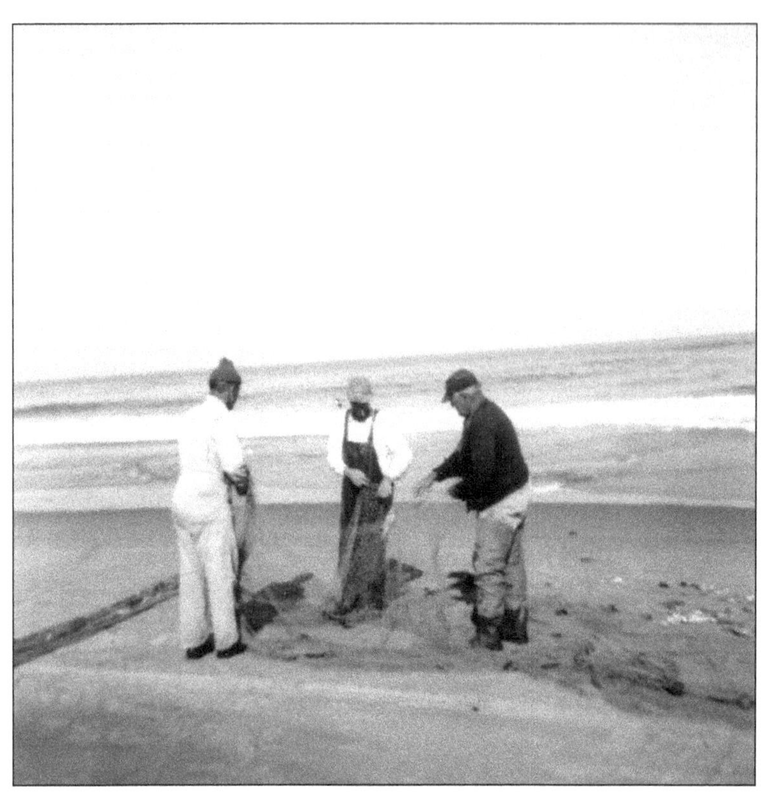

Jethro Jr., Charles Johnson, and Jethro Midgett Sr. fished within a two-mile stretch where they knew there were good sloughs and no shipwreck hangs. Twenty-five years later, coauthor Wayne Gray set a net there and had the 20,000-pound catch of a lifetime. Almost every mesh held a five-pound gray trout. To keep them from spoiling in the sun, they had to be cut out of the net with pocketknives. (Courtesy of Jeffrey G. Midgett.)

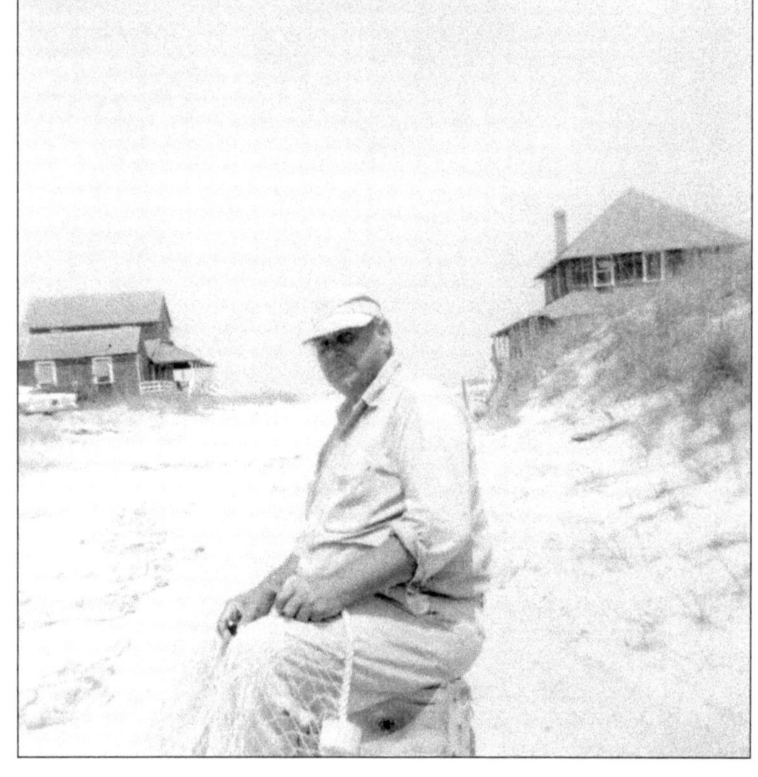

A very talented net mender and hanger, Charlie Beasley spent untold hours sitting on a wooden fish box with a net needle in his hand and his work draped over his lap. The net with cork bobbers stretched out behind him in the Nags Head sand in March 1959. The old cotton webbing needed constant repair, but things changed when nylon, and then strong monofilament, netting was introduced. (Courtesy of Jeffrey G. Midgett.)

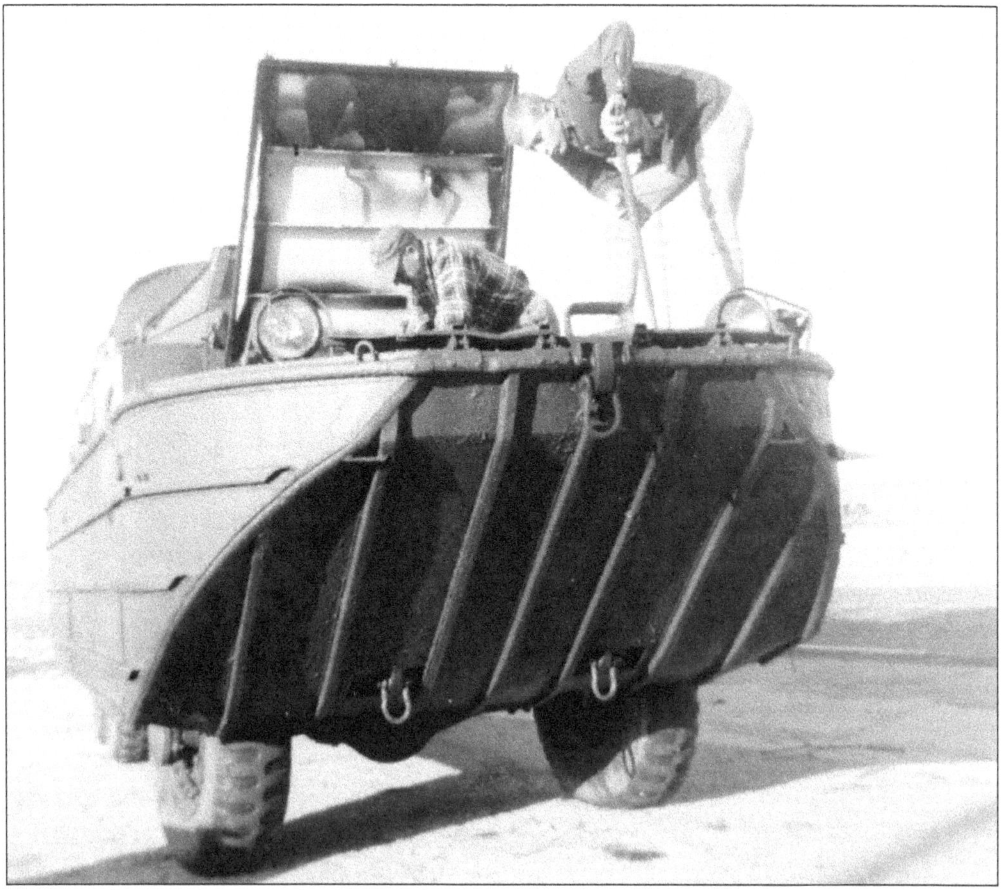

Father and son, Jethro Jr. and Jeffrey Midgett, worked to get the DUKW going. Jethro was known for his extreme strength brought about by a life of hard work. Once, when teenaged Jeffrey got his foot tangled in the rope of an outgoing net, Jethro held the net, pitting his brawn against a forward driving inboard motor long enough for Jeffrey to free himself. (Courtesy of Jeffrey G. Midgett.)

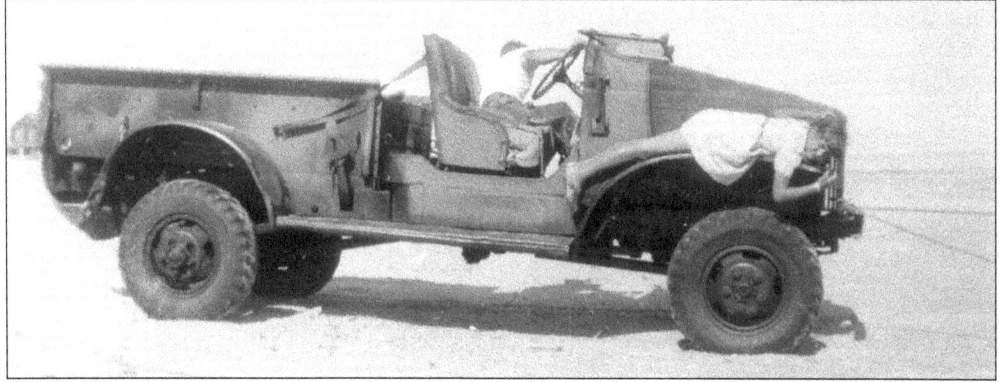

Like a bagged trophy from an African safari, Carmen Pridgen stretched over the fender of her uncle's jeep. Jethro Midgett Jr. paid no mind to her as he pulled the line with the World War II remnant. About 50 years later, when Nellie Myrtle Pridgen died, Carmen had her mother's ashes scattered in the ocean from a dory. The memorial goers gasped as they watched the dory almost capsize. (Courtesy of Jeffrey G. Midgett.)

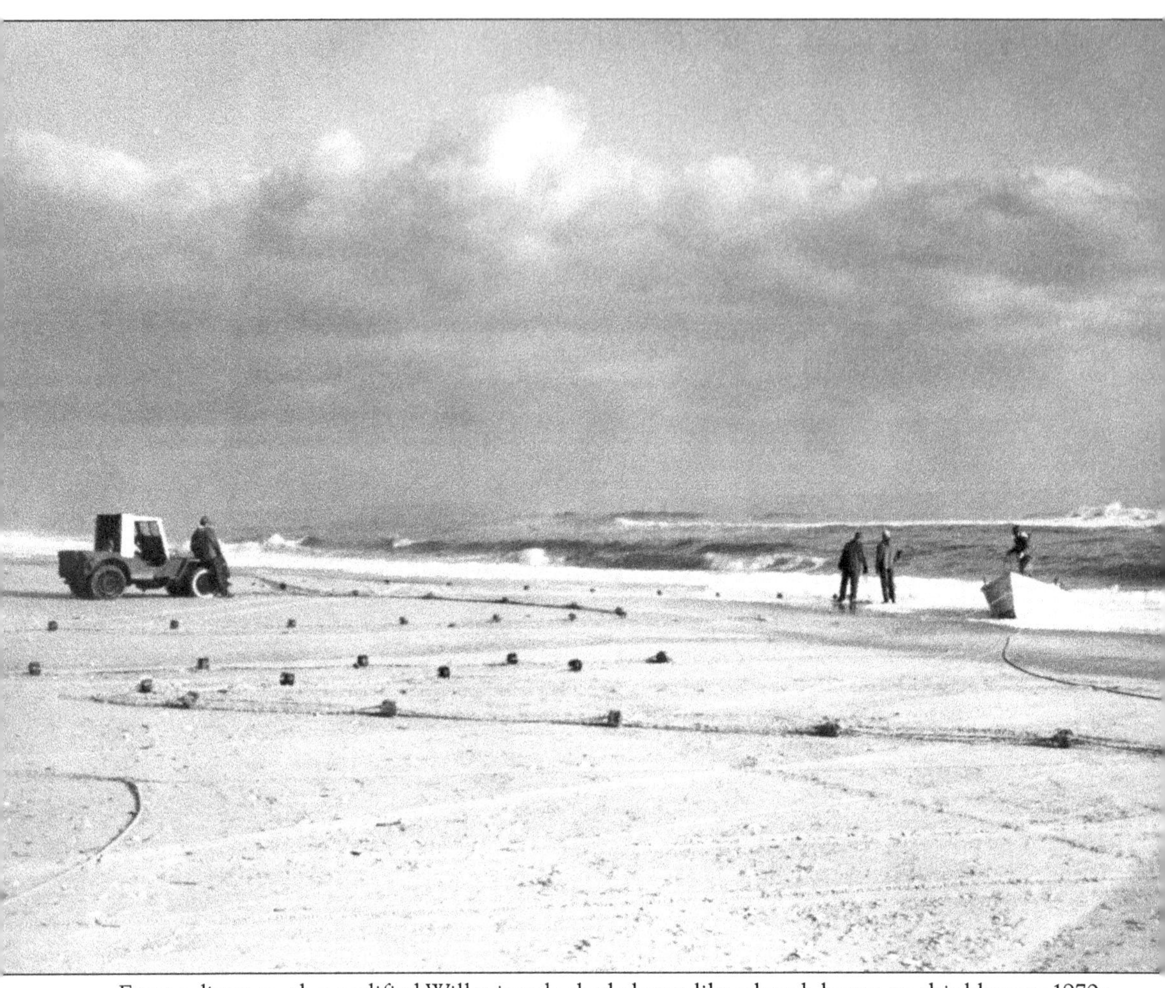

From a distance, the modified Willys jeep looked almost like a beach buggy on this blustery 1970s day. The dory would soon be pulled up on a trailer, the net straightened out, and then loaded back into the dory. The substance that changed a fisherman's lot most significantly was ice. Until ice came to the Outer Banks, fish had to be sold locally or preserved with salt or smoke for later marketing. Before the Civil War, 500-pound blocks of ice were cut out of New England ponds, packed in insulating hay, and loaded in the tightly sealed holds of ice schooners that sailed for Southern shores. Amazingly enough, normal meltage rarely exceeded 10 percent. Cabell Fletcher wrote about the regional icehouses that stored the white gold in *Beneath the Shifting Sands*, saying, "These houses were like two boxes, one within the other and filled with sawdust for insulation. . . . It was only used by fishermen for shipping fish. It was too expensive and people did without." (Courtesy of Jeffrey W. Midgett.)

Pictured in 1955, a barefoot Jimmy Culpepper is working with Jethro Midgett Sr. in oilskins and young Jeffrey Midgett in a sweater. Some locals, like Wayland Baum, harvested ice during rare freezing weather conditions. When the ice began to break up, they would go out in boats and dip out blocks of ice that broke off of frozen swells in the sound. "That was hard work, I tell you," said Baum. (Courtesy of Jeffrey G. Midgett.)

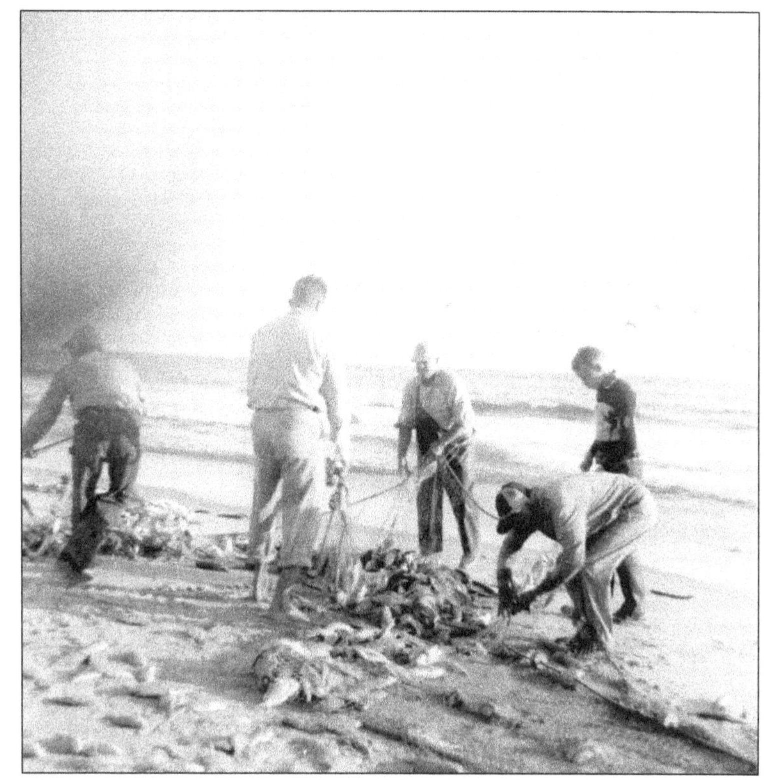

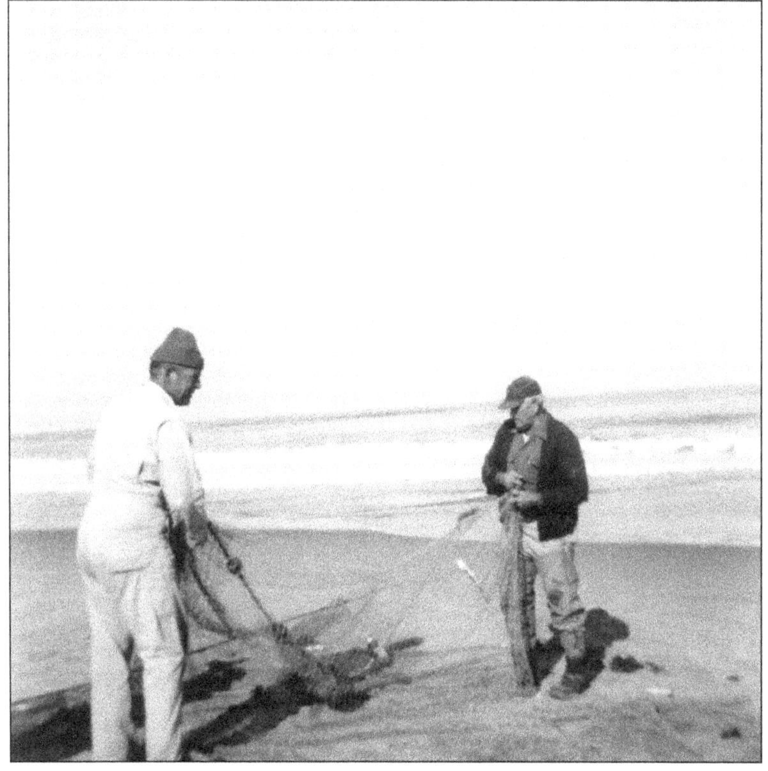

Jethro Jr. and Sr. had their ice delivered from the Dare County Ice and Storage Company, located in Manteo. The 300-pound blocks were scored and broken into six equal pieces. A 50-pound cube could be cautiously dropped into Midgett Seafood's chipper to produce crushed ice to cool down fresh fish. (Courtesy of Jeffrey G. Midgett.)

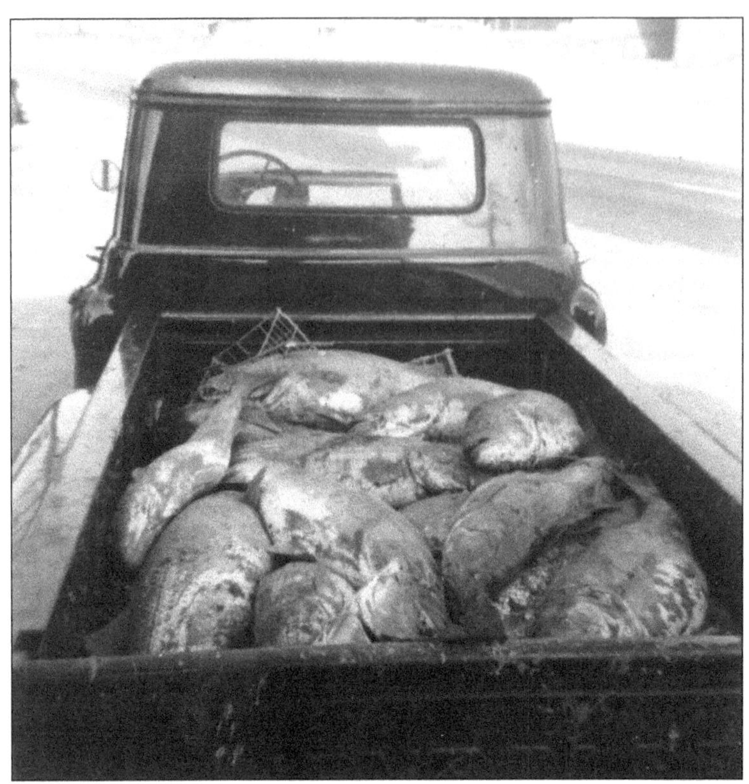

A truck bed of sandy rock was brought up from the beach to Midgett's Seafood in 1956. The striped bass would be dunked in a bathtub of clean water behind the fish house. Big catches in the 1950s and 1960s supplied a growing tourist population with signature Outer Banks seafood. When iced, the overflow could easily be sold to markets in Norfolk or Elizabeth City. (Courtesy of Jeffrey G. Midgett.)

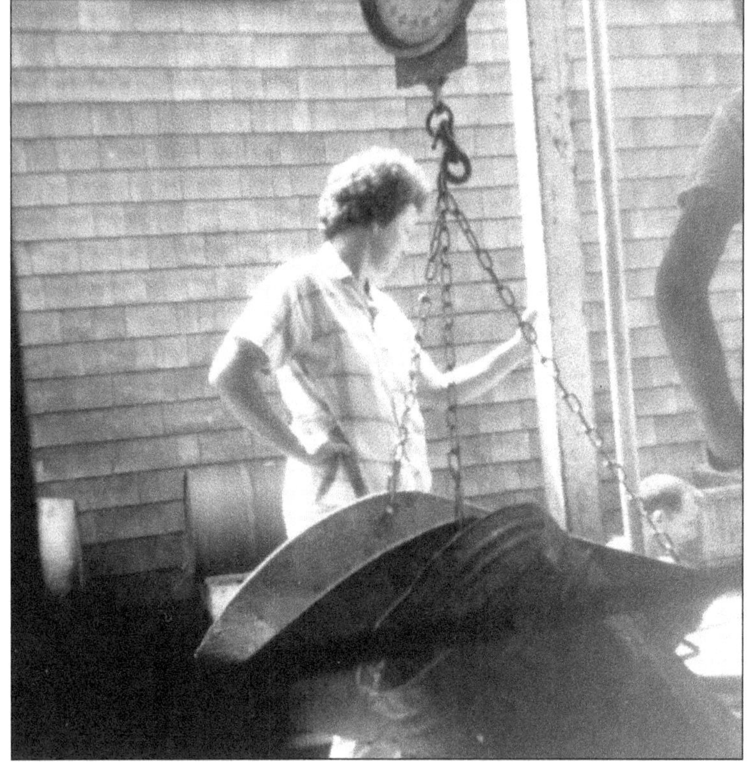

In 1965, Carolyn Johnson waited for the morning's haul to be weighed and tallied. She and her husband, Ivey Johnson Sr., fished a beach rig together. Of her fishing days, she recently said, "I always liked it." Whether it was in a supportive capacity like preparing meals or keeping books or a primary role, like Carolyn played, women were a vital part of the fishing family. (Courtesy of Jeffrey G. Midgett.)

Looking fit and handsome, this photograph of a young Jethro Midgett Jr. wound up on a postcard produced by Graycraft Card Company. The cards were used to promote tourism for the remote island community. Being a fisherman was once idealized as an honorable and free way of life but is now often demeaned as an occupation for bandits by opponents to commercial fishing. (Courtesy of Jeffrey G. Midgett.)

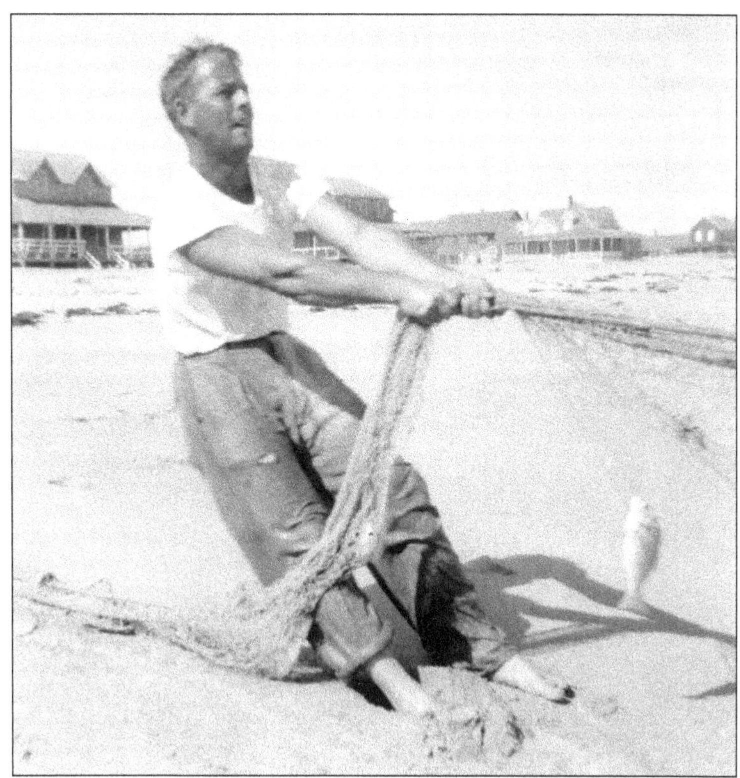

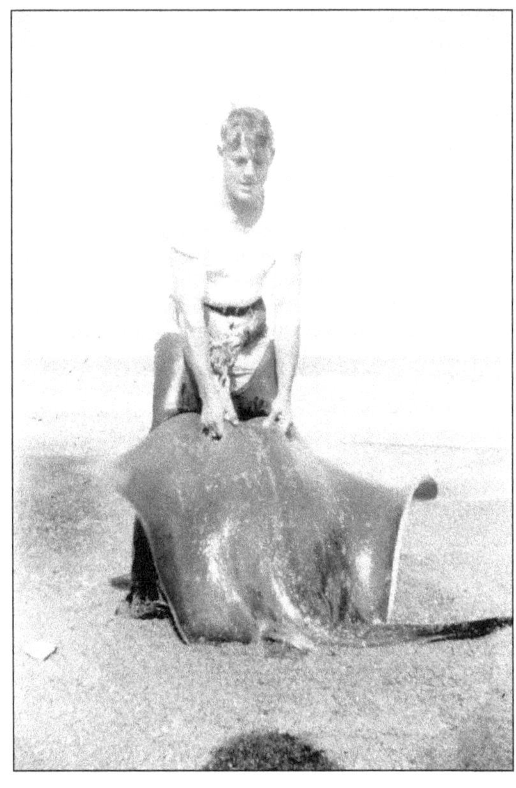

Holding a huge netted stingray by its eyes was an effective way to restrain it. That method did not always work for Jethro Midgett Jr., who was once stung by a sharp tail barb. There was no antidote for the injury, and he was incapacitated for six months while the flesh around the wound first died and then slowly healed. Many watermen have been stung in a similar manner. (Courtesy of Jeffrey G. Midgett.)

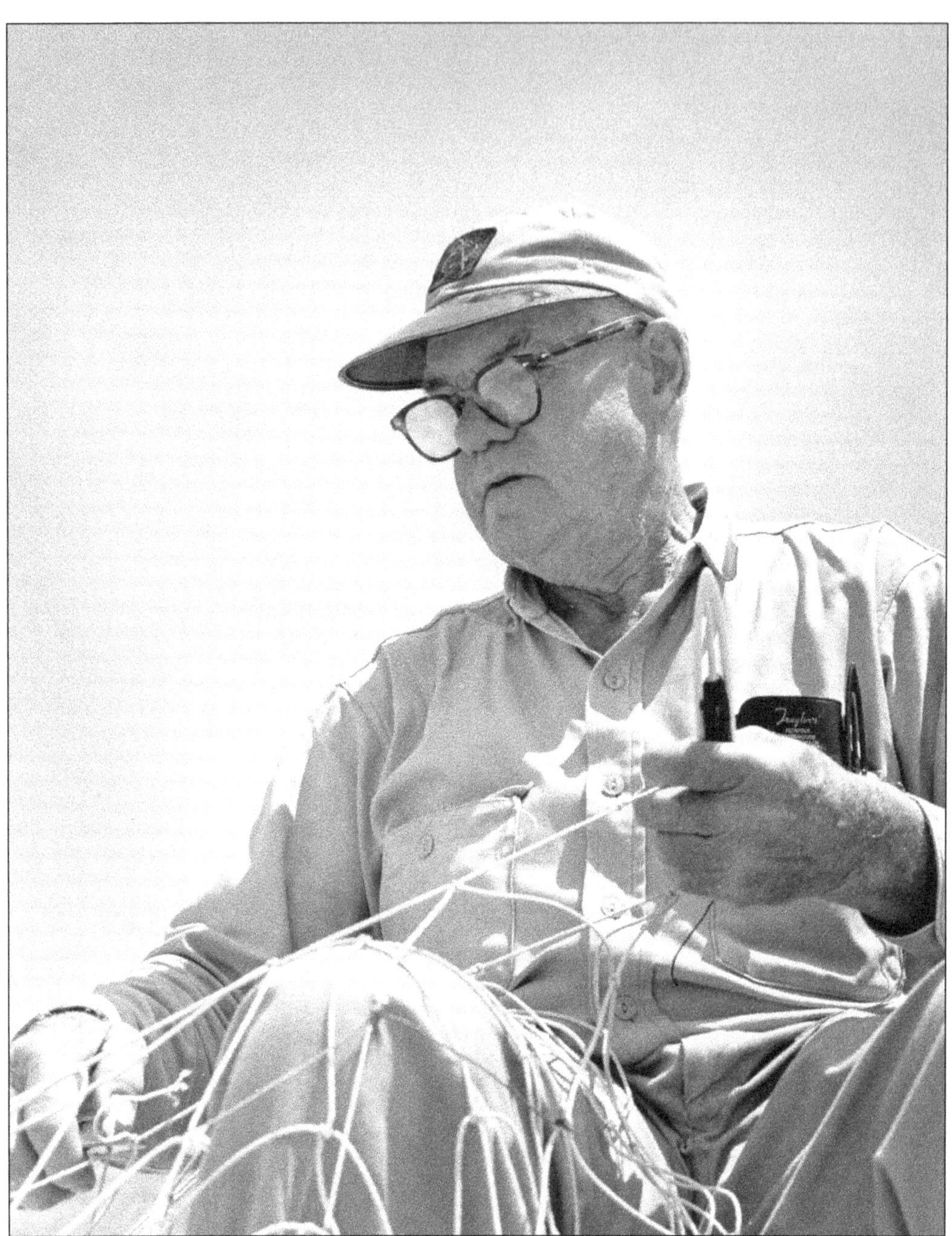

"Old Man Midgett," as his great-nephew and some others called him behind his back, had a wealth of knowledge about the geological make-up of the Outer Banks. Jethro Midgett Sr. believed them to be an elevated sand bar that the ocean needed to sweep over during a storm. He was vehemently opposed to dune construction and the building of the bypass highway, predicting it would act as a dam. Although his rants fell on deaf ears, he was proven right in his assessment. During the Ash Wednesday Storm of 1962, a great deal of damage occurred when ocean water did break through the "dam" and created a fast moving river that took everything in its path. His vocation as commercial fisherman made him an avid weather forecaster and observer of nature. (Courtesy of Jeffrey G. Midgett, photograph by Gladys Young.)

Four

Beyond Subsistence

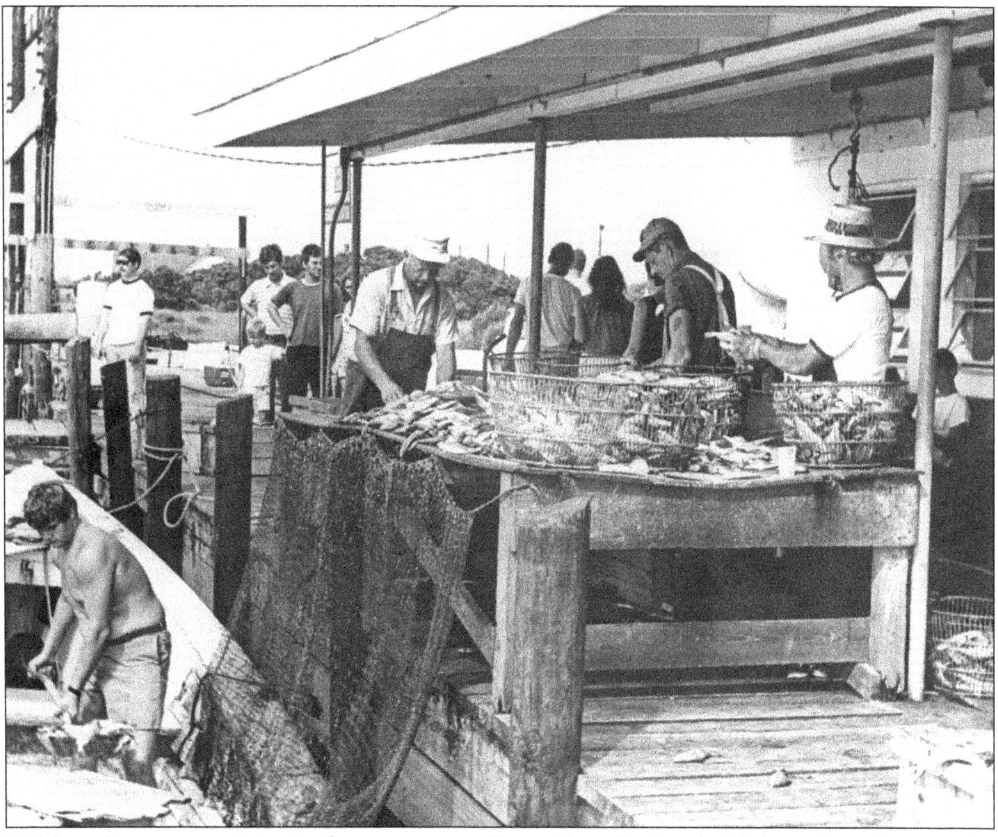

Holding the worst position in the crew, young Steve Daniels shoveled fish up from down in the boat. The net attached to the wooden table kept any fish from spilling overboard. From left to right, the next two men, Charles and Robert Daniels, culled through the catch. Wanchese long net rigs went as far as Hatteras to seek sound catches on well-known shoals. (Courtesy of Aycock Brown Collection, OBHC.)

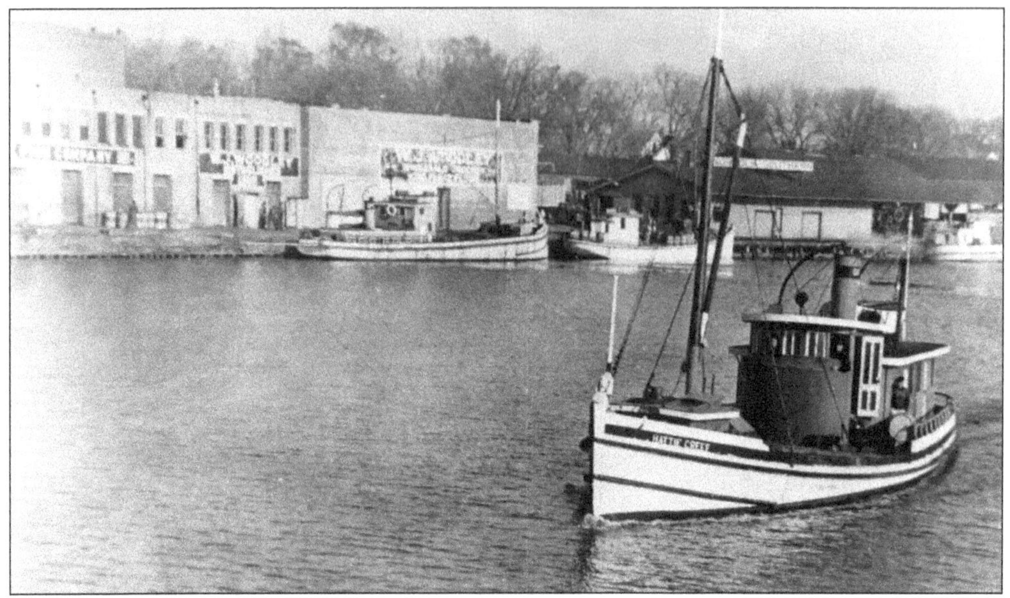

Businessman E.R. Daniels owned a fleet of sailing vessels that provided fishermen a way to ship their seafood out of the county. Stations located at Mashoes, Manns Harbor, Stumpy Point, Wanchese, Rodanthe, Avon, and Hatteras were serviced by boats, like the *Hattie Creef*, that transported everything from mail and passengers to ice and fish. (Courtesy of OBHC.)

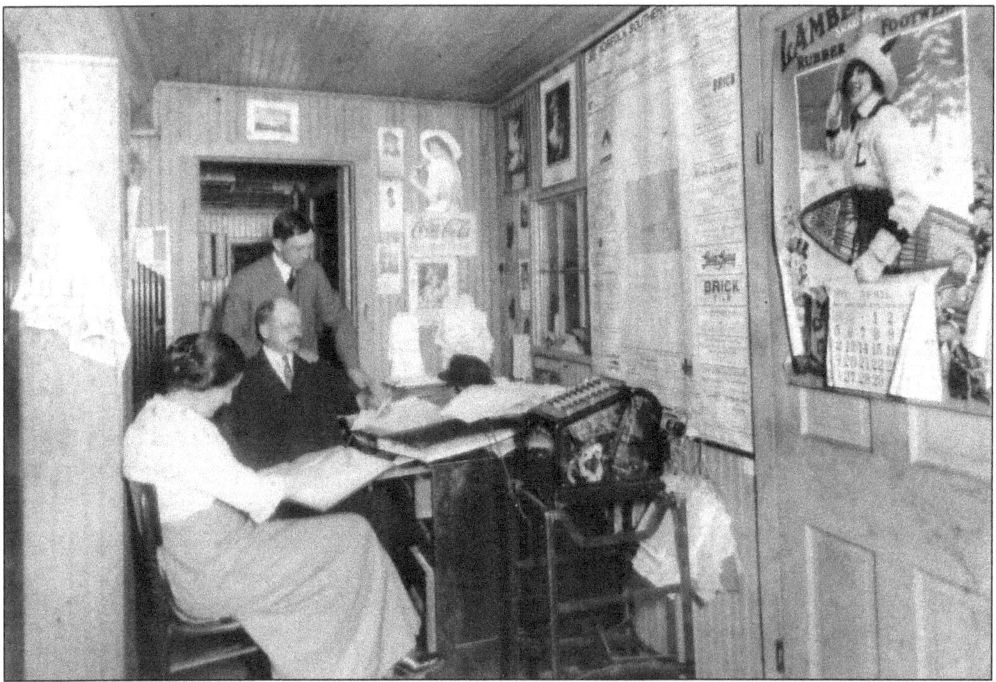

Globe Fish Company's main office was located in Elizabeth City, and was operated by Daniels's well-educated descendants. This 1914 work area featured a window from which to keep watch on the fish room and docks, a "Norfolk Southern Railroad Official Freight Shipping Guide" on the wall, a glass-sided adding machine, and a safe behind the secretary. (Courtesy of Museum of the Albemarle.)

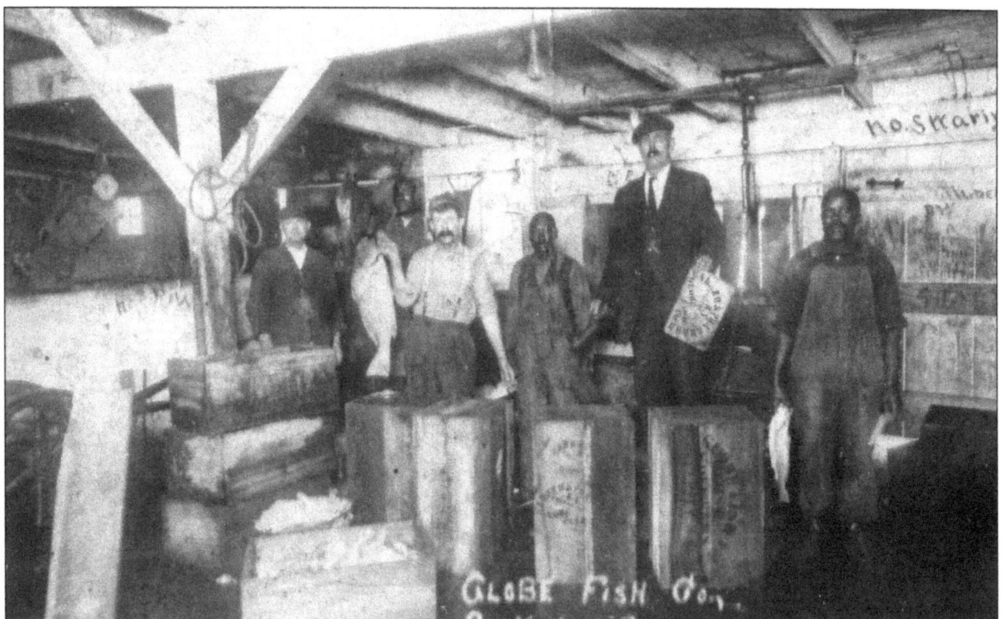

"No Swaring" was the rule in the Globe Fish Company packing room. The man in the tie held a brass stencil in one hand and an ink roller in the other. These particular wooden fish boxes were stamped with the arcing name of a fish dealer at Fulton Fish Market in New York City where they will be shipped. At the bottom of the half-circle was "Globe Fish Company," the sender. (Courtesy of Museum of the Albemarle.)

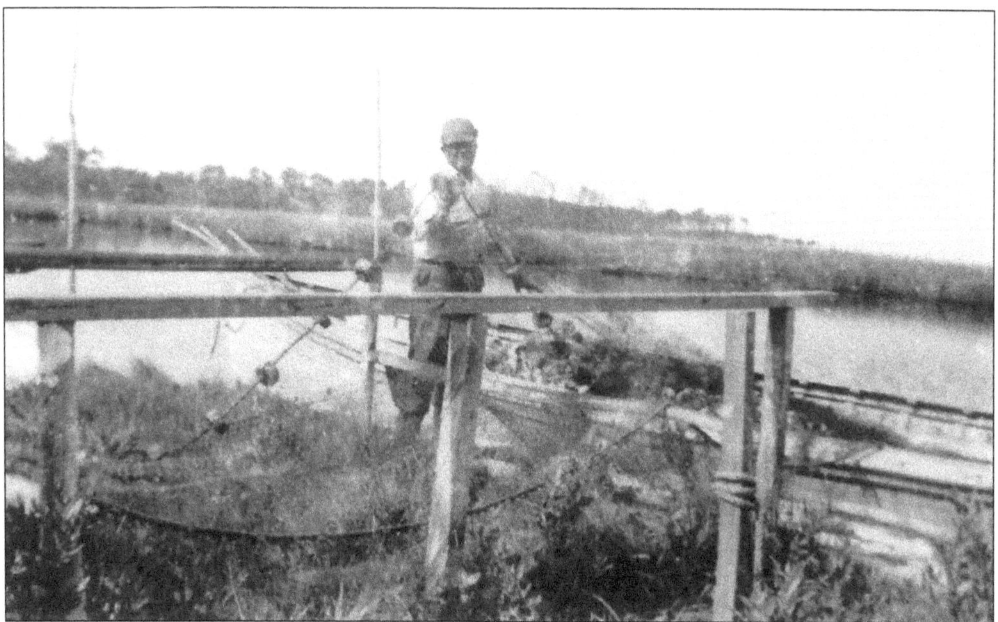

Roanoke Island fishermen used the marshland as a staging area for their operations. Lloyd Wescott pulled a net from his boat to drape over wooden structures that held up the mesh as it dried in the sun. At certain times of the year, it was heavily tarred to help preserve and strengthen the cotton. A net house was usually nearby to store the nets out of the sun when not in use. (Courtesy of Lou Ellen Quinn.)

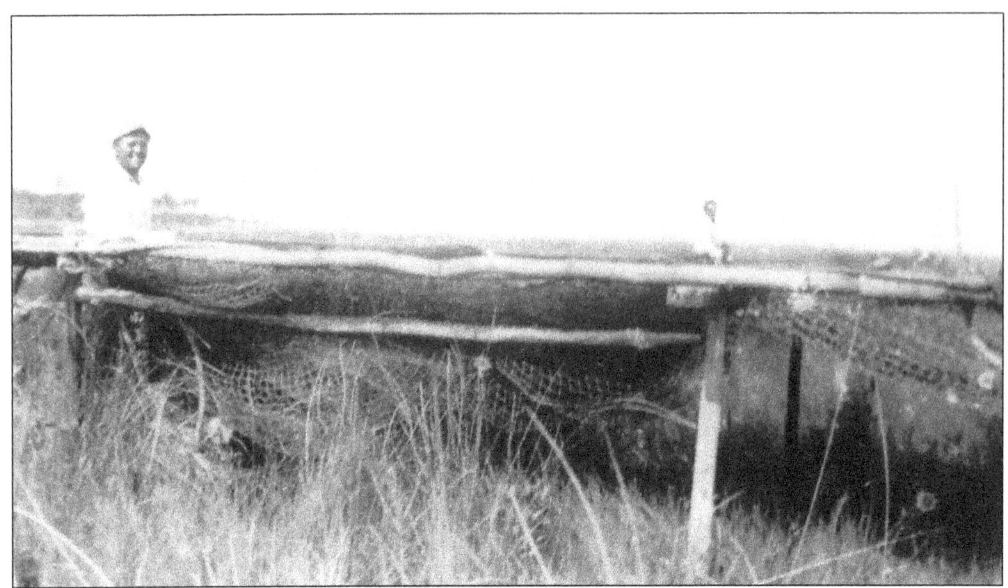

Thirty-year-old Ab Wescott was racking the nets for his father in 1940. Long netters would have the strenuous job of hauling in almost a mile of heavy net, slowly reducing a large circle to a small bunt. It was no wonder that Lloyd's sons, Ab, Fred, and Roy, borrowed money from their father to embark on a new business venture in 1945, Manteo Furniture Company. (Courtesy of Lou Ellen Quinn.)

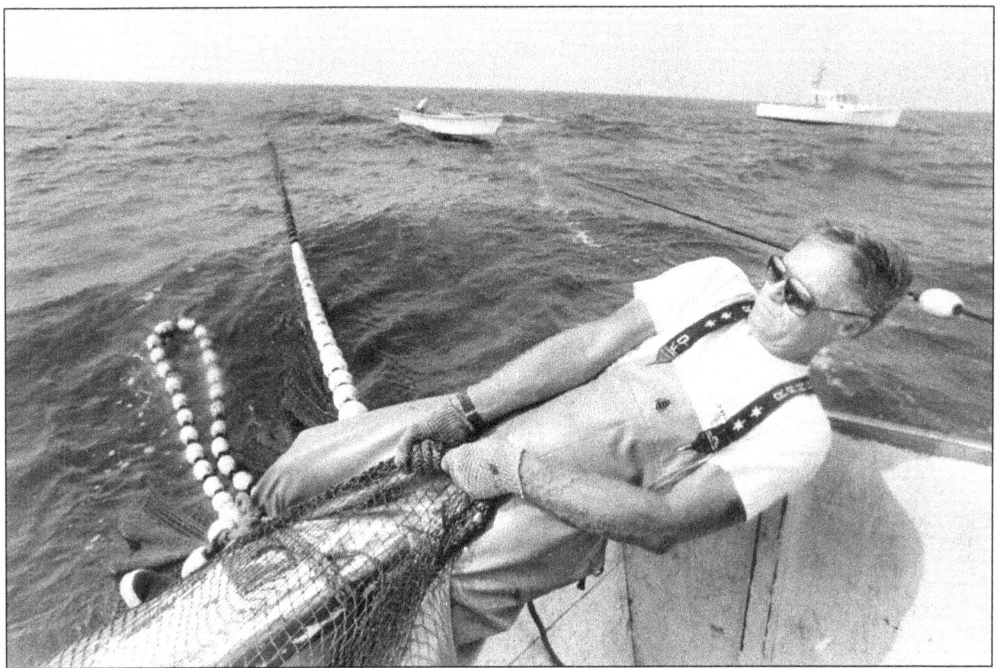

Long netter Carter Melton gave it all he had as he heaved 2,000 yards of net in 1986. The cork floats attached to the line at the top are readily seen. Underneath the water, the line at the bottom is held down with lead weights. He most likely left his home before daybreak and rode in the cabin boat in the background that towed the other two workboats. (Courtesy of Drew Wilson Collection, OBHC.)

Once the fish were impounded in the bunt, the only way to get them out was with long dip nets. This group used a rope hoist to help with the heavy weight. Men of old would dip out the fish and throw them over their shoulders into the fish boat. Long netting continued through March and April. Around the first of May, spring runs of rock, shad, and herring ceased. (Courtesy of Drew Wilson Collection, OBHC.)

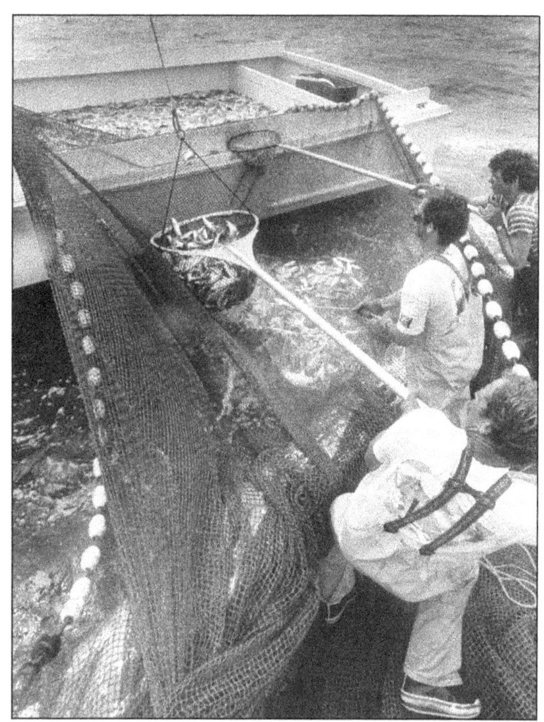

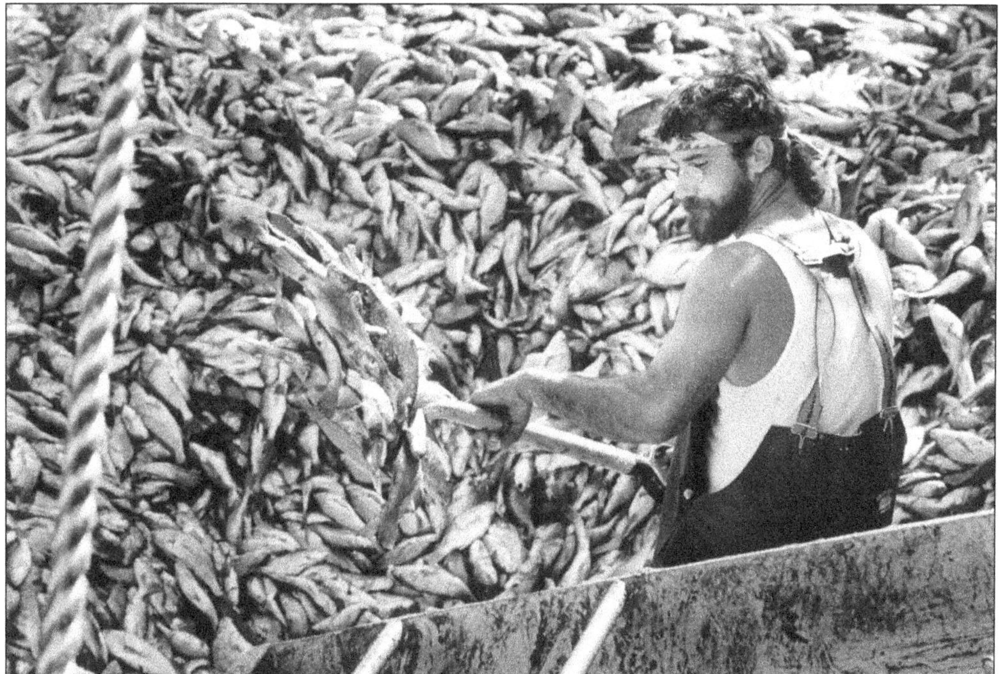

It appears as if Etheridge Seafood Company employee Keith Samford was single-handedly trying to tackle a mountain of croakers with a small shovel. Large catches of fish brought about an all-hands-on-deck condition with calls for extra help. In the 1970s, boys were even allowed to check out of Manteo High School early to go and work in the fish houses. (Courtesy of Drew Wilson Collection, OBHC.)

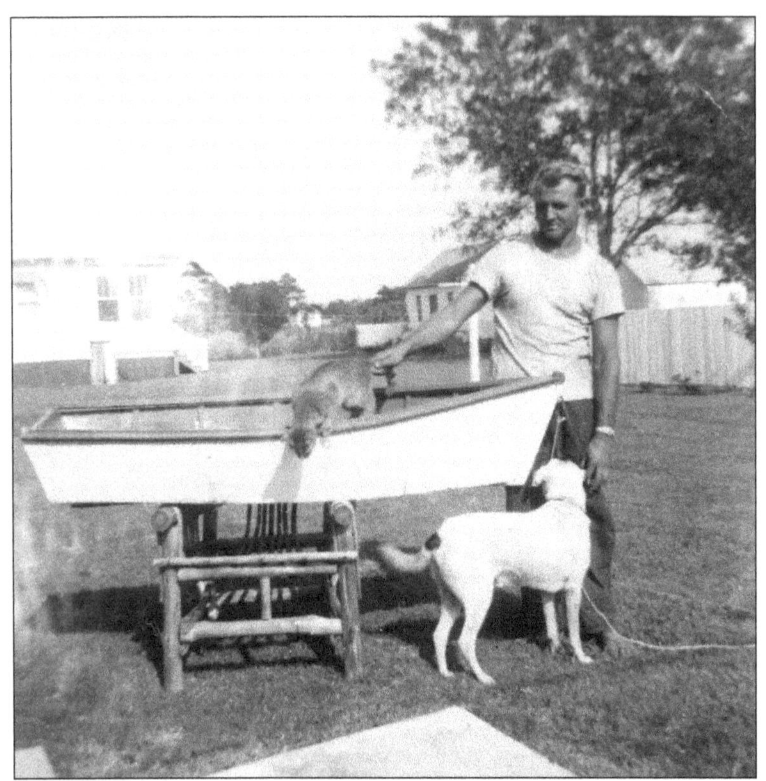

Erbe Gallop received feline and canine approval for the tiny boat he built with his brother in the 1950s. While fishermen were benefitting from better equipment and wider product distribution, there were still many months of sparse work due to seasonal weather. Many of the men turned to backyard boatbuilding, constructing vessels for themselves and others. (Courtesy of Tex Gallop.)

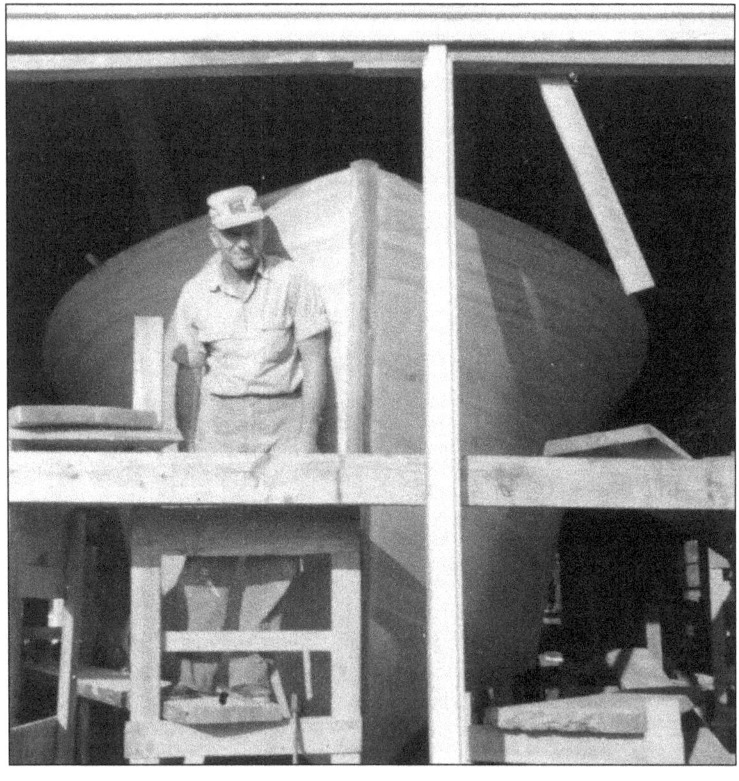

When not fishing his nets, Hubert Tillett worked with legendary boatbuilders Warren O'Neal and Buddy Davis. The solitude of being on the water and not answering to a boss are factors that appeal to all fishermen. The trade-off, however, is a certain amount of job insecurity. Paid insurance, steady hours, and a weekly paycheck lure many men away from their heritage and vocation. (Courtesy of Stuart Bell.)

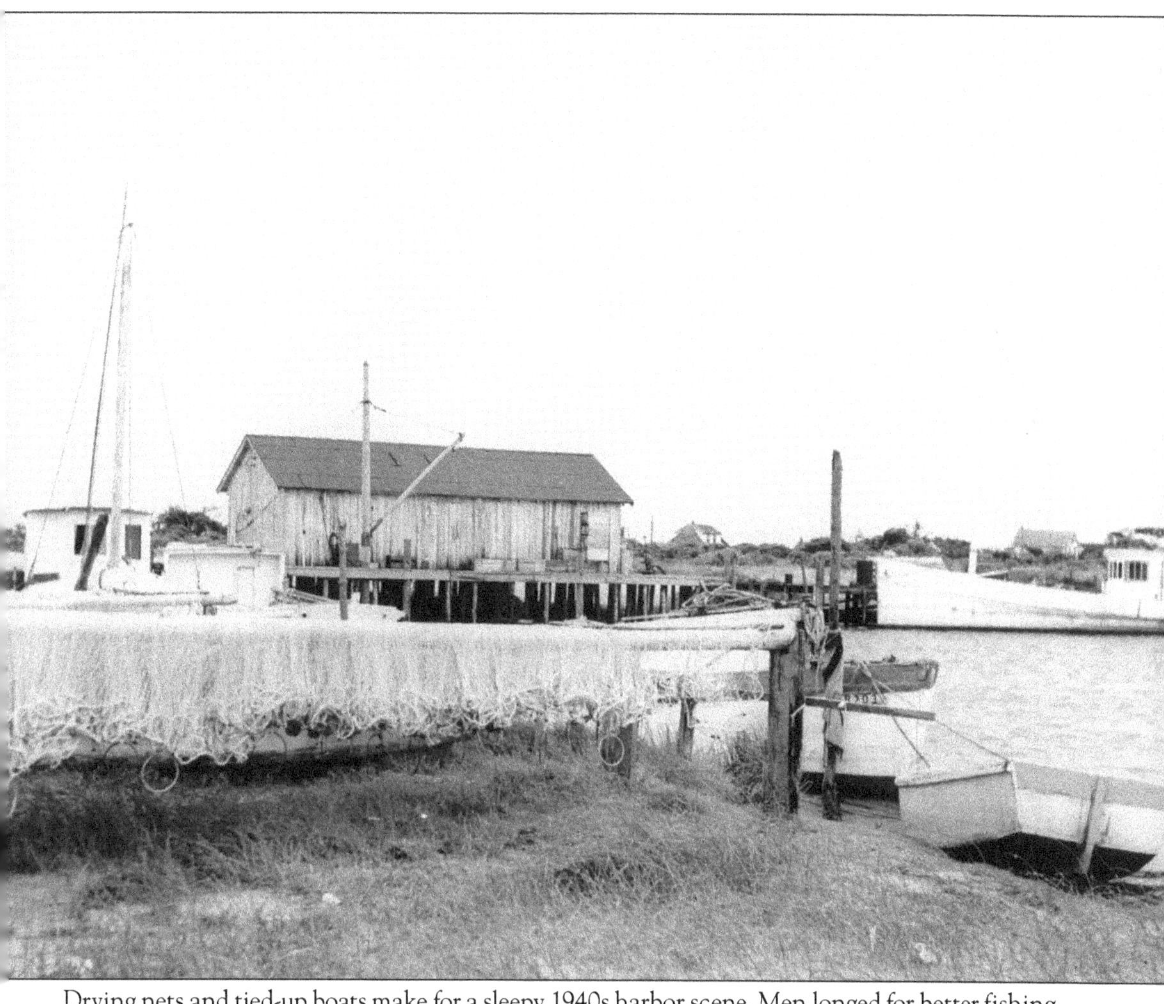

Drying nets and tied-up boats make for a sleepy 1940s harbor scene. Men longed for better fishing opportunities and got them after World War II. A combination of circumstances worked to put commercial fishermen in the best position they had ever known. A hyperbolic radio system for long-range navigation, called Loran for short, was developed during the war and became available to civilians shortly after. They could safely venture farther out to untapped grounds, formerly unattainable to them. Fishermen also took advantage of retired war vessels that were coming on the market. With creativity and elbow grease, the boats, already proven to be seaworthy, were repurposed into fishing trawlers. Some men dabbled in the recreational fishing industry, stripping down their boats for charter fishing in the Gulf Stream in the summer, and rigging up to return to commercial fishing in the off-season. (Courtesy of Aycock Brown Collection; OBHC.)

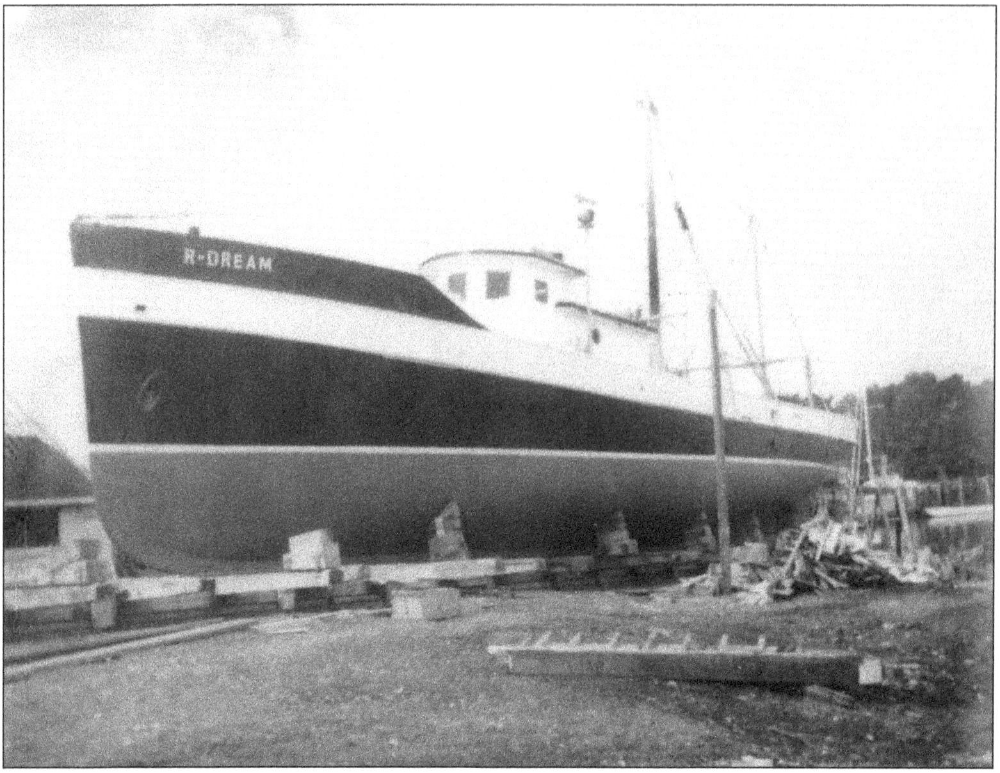

The *R-Dream* began as a handsomely appointed cruiser on the Great Lakes. During her second career, trawler *R-Dream* took out legislators in 1975 to show them how a catch was brought in. Rough conditions greatly lengthened the trip, and the group missed its hot shore-side lunch. The resourceful captain, Dennie Daniels, passed around a box of Frosted Flakes to assuage their hunger. (Courtesy of Robin Daniels Holt.)

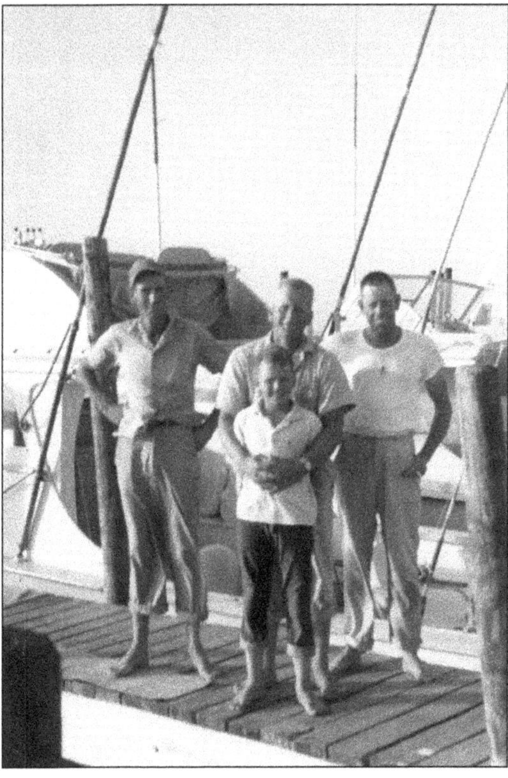

Posing on the Hatteras Docks after a day of long netting, from left to right, Dennie, Steve, Charles, and Robert Daniels looked like natives in rolled-up trousers. Dennie took the *R-Dream* and his family to Greenport, Long Island, New York, for many summers to shrimp. After taunting from local kids, his daughters quickly learned to curb their Outer Banks' "hoi-toider" accents. (Courtesy of Robin Daniels Holt.)

Rex Etheridge's brogue was so thick that new acquaintances sometimes thought he was from England. Outer Bankers have an accent that linguists think dates back to Elizabethan English. Using their own vocabulary, natives might say their boats were "mommocked," or roughed-up, during a storm. The dialect has become less pronounced as the area has become less isolated. (Courtesy of Rex Dea Etheridge.)

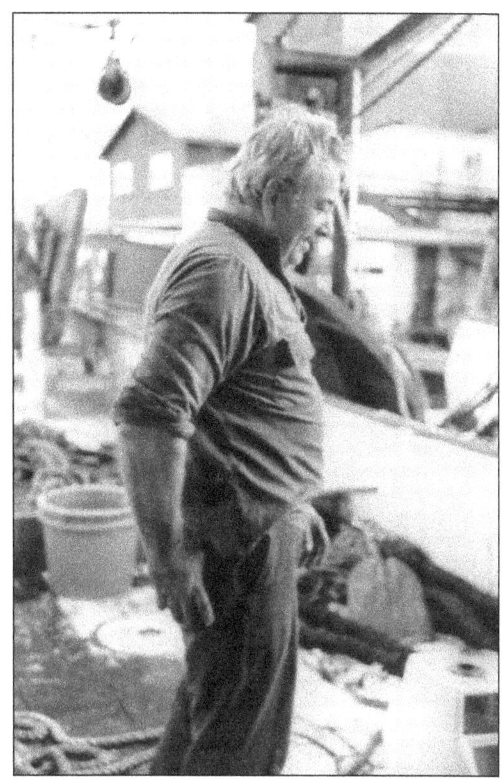

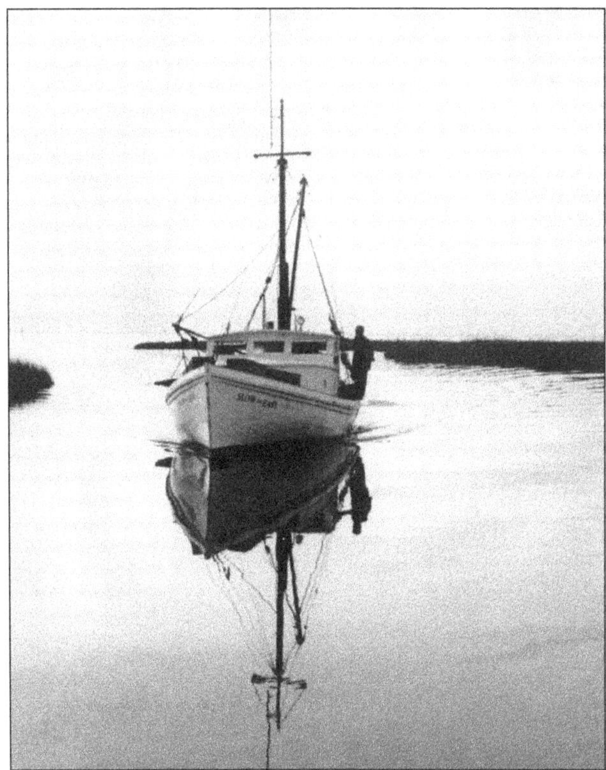

In the days of Fred Basnight and the *Slow an Easy*, fishermen had to be their own weathermen, relying on the barometer and taking action if it dropped quickly. Watermen paid attention to wind speed and direction, the flight of birds, rings around the moon, tidal ebbs and flows, and the color of the sunset. Their very lives depended on making accurate forecasts. (Courtesy of Aycock Brown Collection, OHBC.)

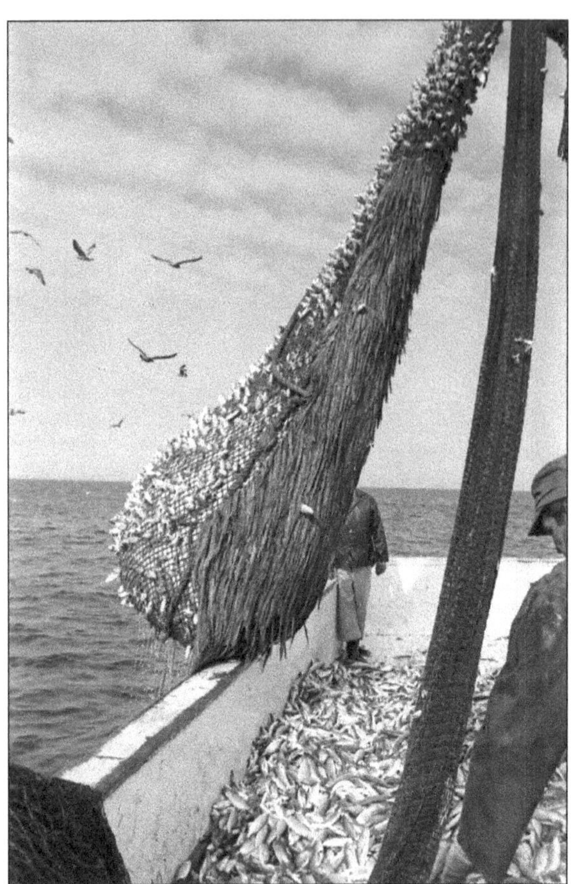

An adrenaline rush comes with hoisting aboard a full net bag. Merle Jean Daniels once equated trawl fishing with gambling. She explained that it could be an addiction for financially desperate fishermen who kept thinking their next big catch was just around the corner. She encouraged her husband to give up his trawl boat before it bankrupted them. (Courtesy of Bruce Roberts Collection, OBHC.)

With the release of a rope, the contents spilled out with a slippery sounding "swhoosh.' Going past the three-mile limit into federal waters brought about better harvests, but more red tape. Logbooks that were once voluntary became an albatross with demands for every i to be dotted and every t to be crossed. (Courtesy of Bruce Roberts Collection, OBHC.)

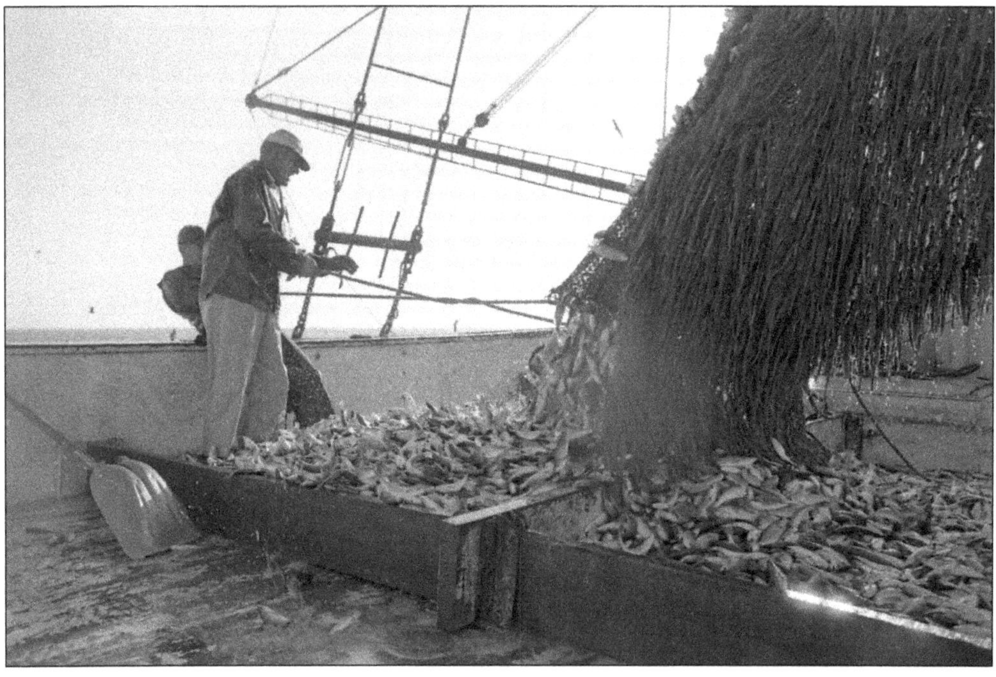

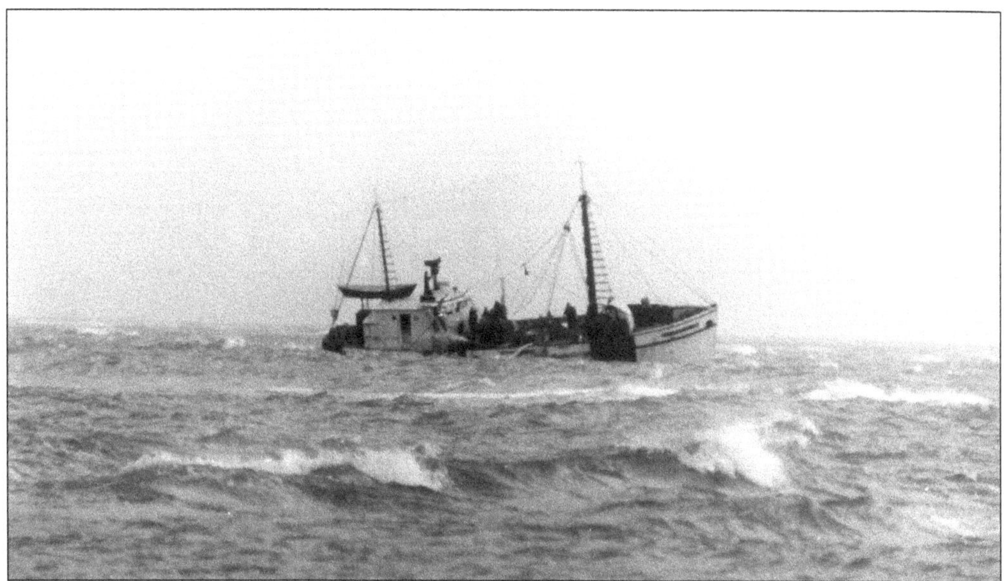

Many trawlers, like 85-foot-long *Sarah J*, have succumbed to the shifting shoals and unpredictable waves of Oregon Inlet. The inlet opened after a violent 1846 hurricane cut a watery path between ocean and sound, separating Hatteras Island from the rest of the northern Outer Banks. Ever since, fishermen prayerfully pick their way through the capricious gateway. (Courtesy of NPS, CHNS.)

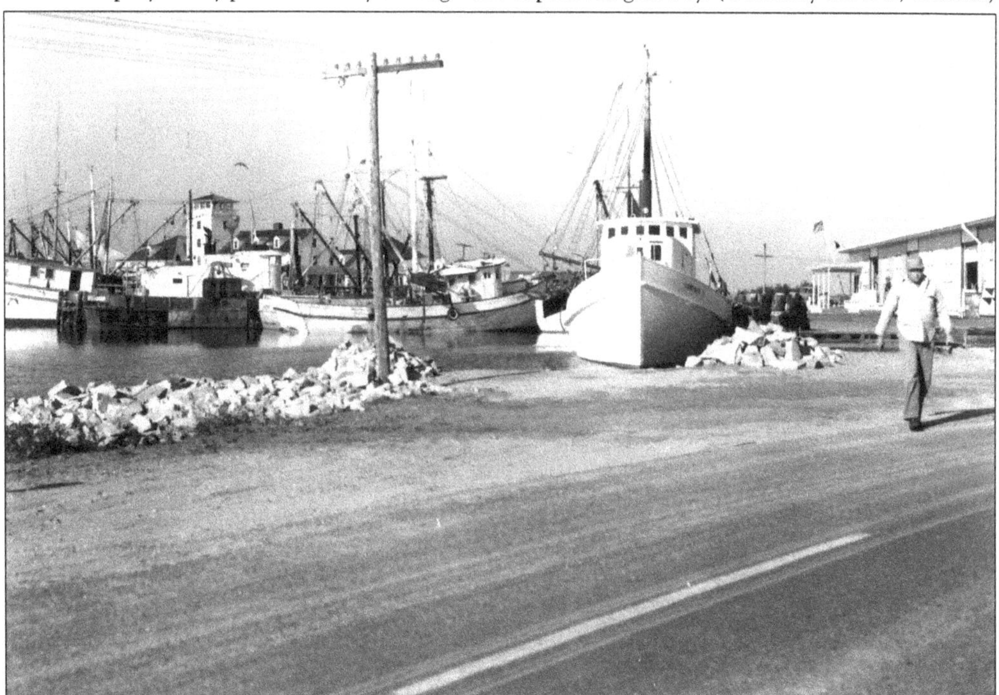

A high tide and an optical illusion make one think this fishing trawler it about to head down Highway 12 in Ocracoke. Fishermen know Hatteras Inlet is as untrustworthy as Oregon Inlet, and indeed, both waterways were formed by the same storm. To the south, deep and navigable Ocracoke Inlet may have been used in 1585 by the colonists on their way to Roanoke Island. (Courtesy of NPS; CHNS.)

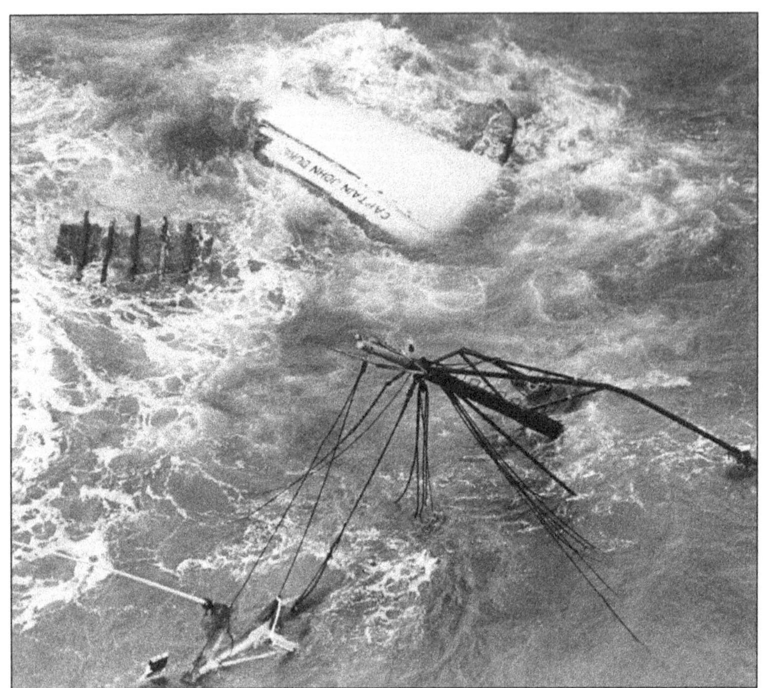

Inlets were not the only treacherous places for trawlers. In 1989, the *Captain John Duke* lost power and foundered on Diamond Shoals, two miles due east of the Cape Hatteras Lighthouse. Four crewmen were rescued by a fellow trawl boat captain, as often happened when a boat sank before the Coast Guard could even be radioed. (Courtesy of Drew Wilson Collection, OBHC.)

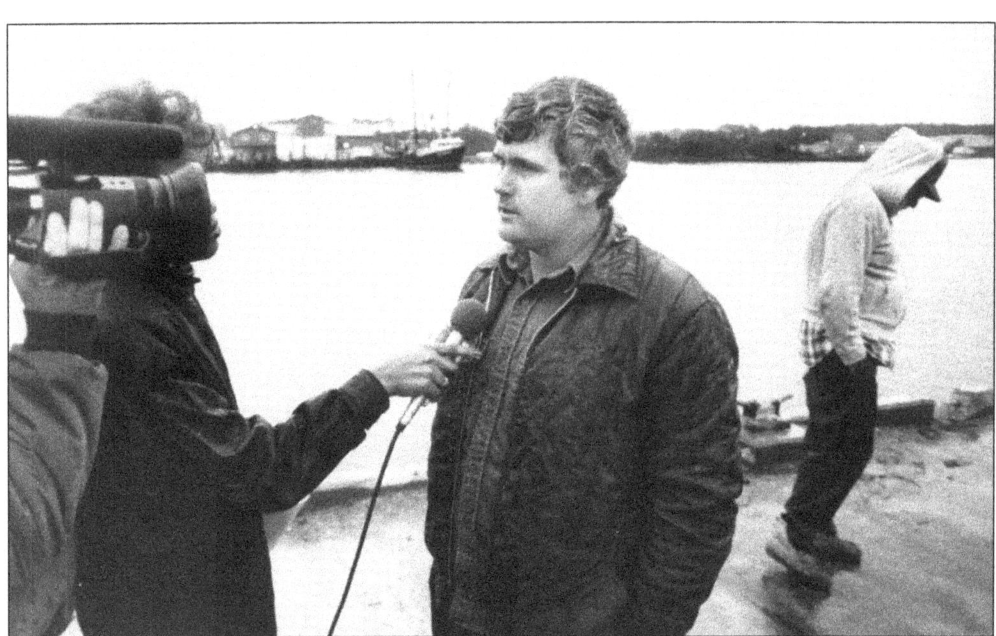

Billy Carl Tillett, unaccustomed to the limelight, found himself as a spokesman for building jetties at Oregon Inlet, making a plea for their construction to a television crew in 1987. The jetties were promised by the federal government decades earlier but were never undertaken because of litigations by environmental groups, National Fish and Wildlife, and the National Park Service. (Courtesy of Drew Wilson Collection, OBHC.)

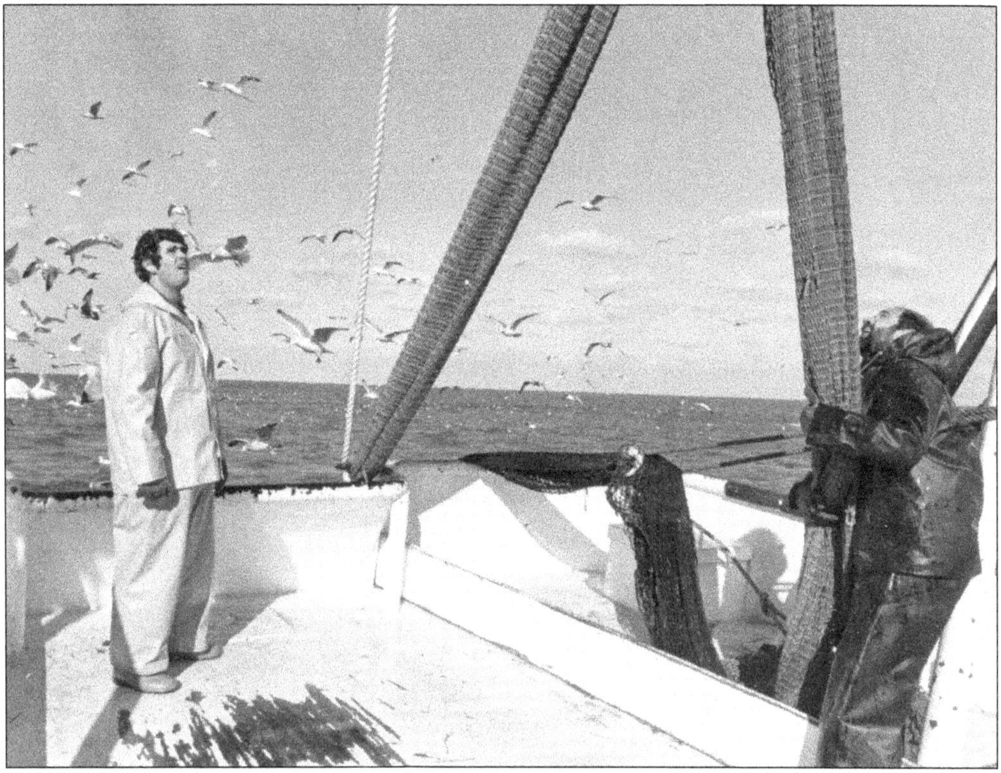

Tillett looked skyward to check his net. Despite constant dredging, the situation at the inlet steadily worsened. Endless political appeals were disregarded, and the dream for stabilization has all but died. Today's fishermen have to concede that the strict regulations that are currently in place are so restrictive that it almost makes the jetties a moot issue. (Courtesy of Billy Carl Tillett, photograph by J. Foster Scott.)

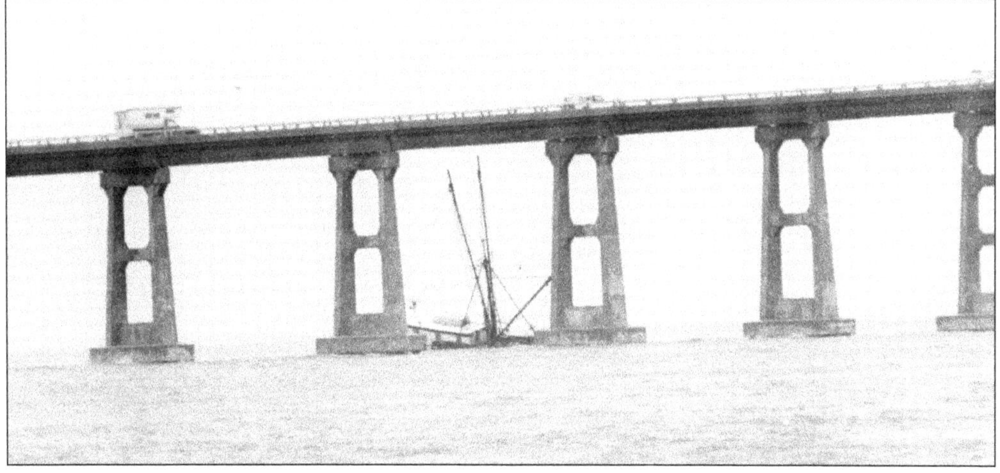

Vehicle occupants whizzing along seemed oblivious to the misfortune manifested below. Toby Tillett, who ran ferries across Oregon Inlet for 25 years, went with a coalition to Washington, DC, in the late 1950s to discuss the building of the bridge. He warned Congressman Herbert C. Bonner that sand would build up around the pilings and shoal up the inlet. (Courtesy of Aycock Brown Collection; OBHC.)

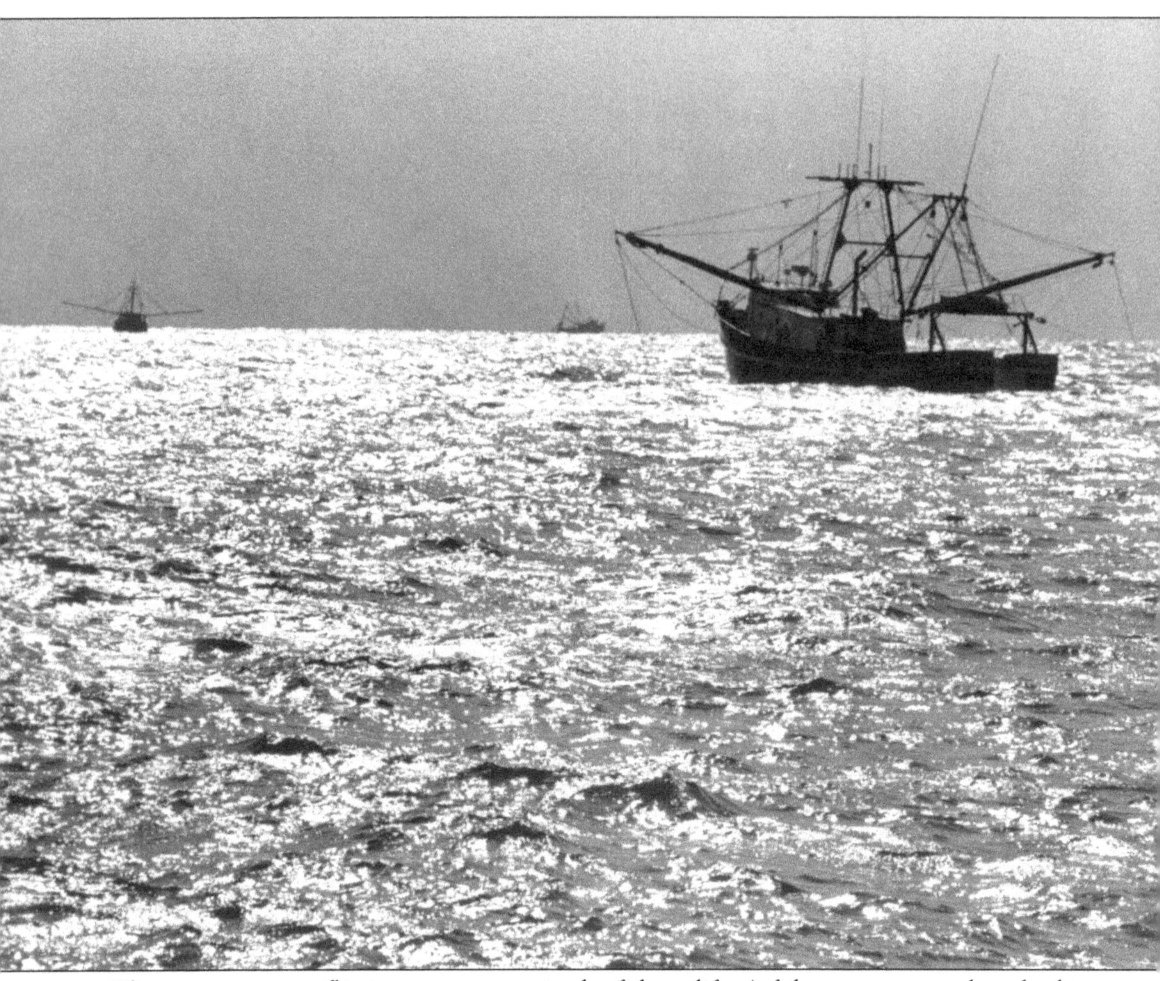

There are many conflicting components in the fishing life. A fisherman can see breathtaking scenes of beauty and terrifying storms of destruction. He can experience a sailor's life at sea with his crew as kin and the longing to be at home with his own family. He can feel the freedom of being his own man and the burden of carrying responsibilities alone. A fisherman needs to have a respect for the laws of nature, as well as the laws of man, whether he agrees with the regulations or not. Sometimes he is seen as opportunistic and anti-conservationist, when, in fact, no one knows better than one who works on the water that it is to his advantage to keep estuaries healthy and fish populations stabile. (Courtesy of Aycock Brown Collection, OBHC.)

Five

Stewards of the Sea

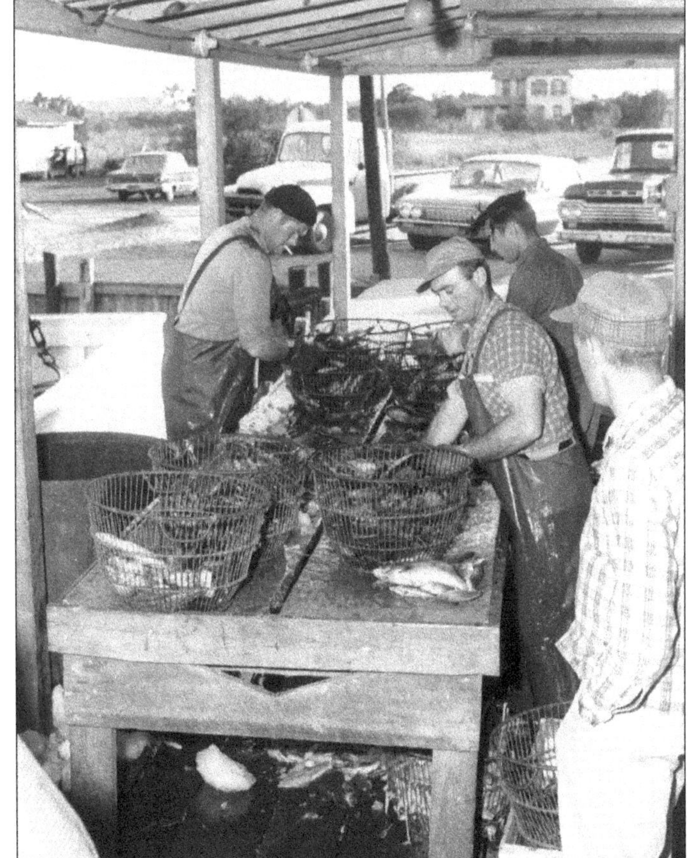

These 1950s workers sort species in wire baskets before icing them down in large wooden boxes. Whenever boats arrived at the Wanchese docks, someone looking for work could easily pick up a job. Young people, who came to live the idyllic beach life, found themselves broke, packing fish in winter when their seasonal occupations as servers, bartenders, or lifeguards ended. (Courtesy of Aycock Brown Collection, OBHC.)

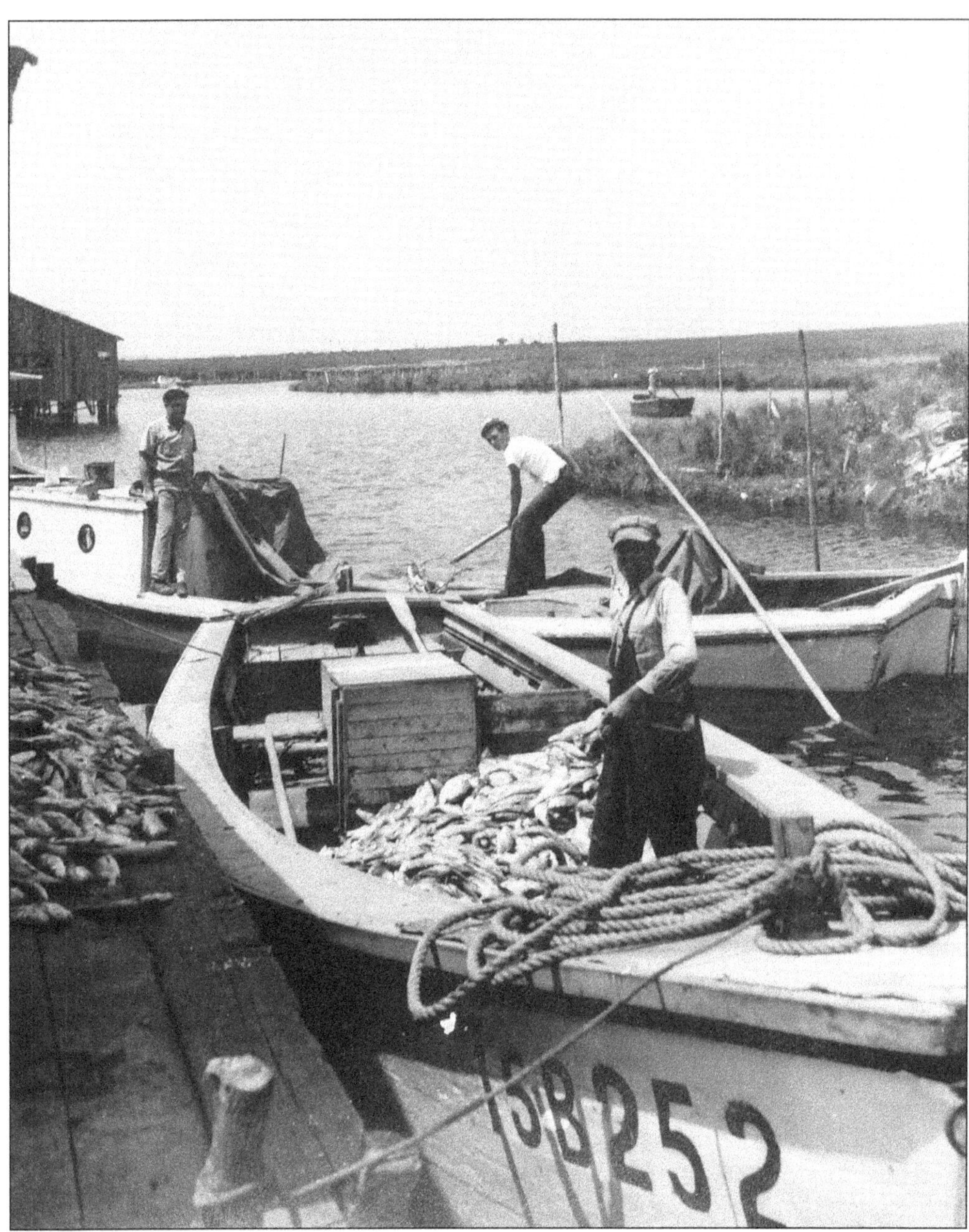

A gentleman through and through, George Albert Daniels, pictured in the foreground, unloaded what will total 10,200 pounds of croakers. In the background, Van Buren Gray and his son Van Jr. wait their turn to offload at Willie Etheridge Sr.'s fish house. In 1945, Daniels came home just in time to take advantage of what was later dubbed "the croaker season." A few years earlier, during World War II, he moved his family to Elizabeth City for work in the shipyard. The boat he stood in originally had a cabin that Daniels elongated. To save money, the family gave up their city apartment and lived aboard the boat next to the shipyard. He returned to fishing after the war, running a long net rig with his brother Arnold. To increase profits, they bought a fish truck that Arnold's wife, Miriam, drove to markets in Philadelphia and New York. (Courtesy of the *Daily Advance*, photograph by Ben Dixon MacNeill.)

Mary Francis Gallop welcomed home her husband, Erbe, and celebrated his good fortune in 1954. Until the 1970s, Wanchese housewives would simply go to Mill Landing for free fish to cook for their suppers. It was a generous tradition, and no one thought it was odd that fishermen never asked for payment and housewives never offered to pay. (Courtesy of Aycock Brown Collection, OBHC.)

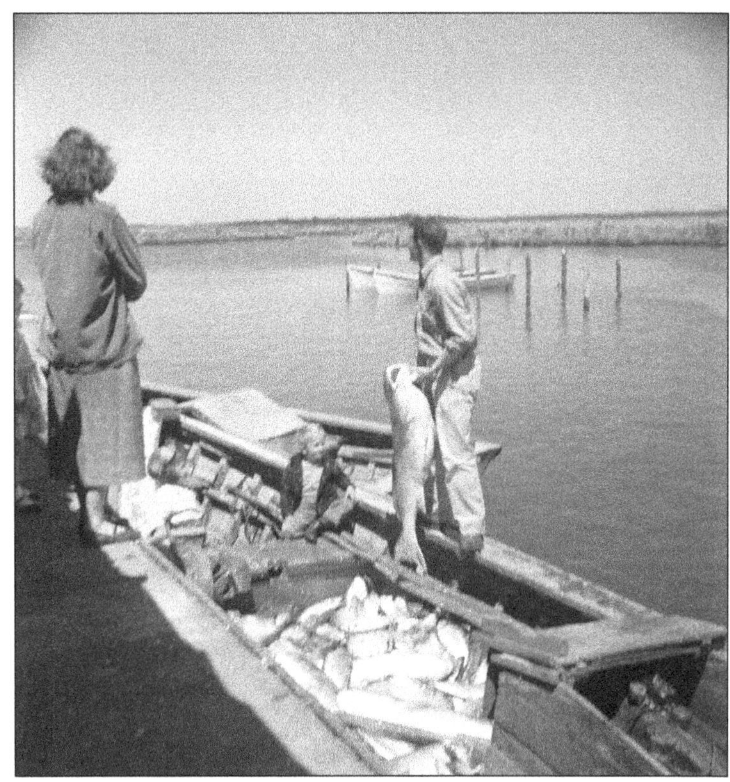

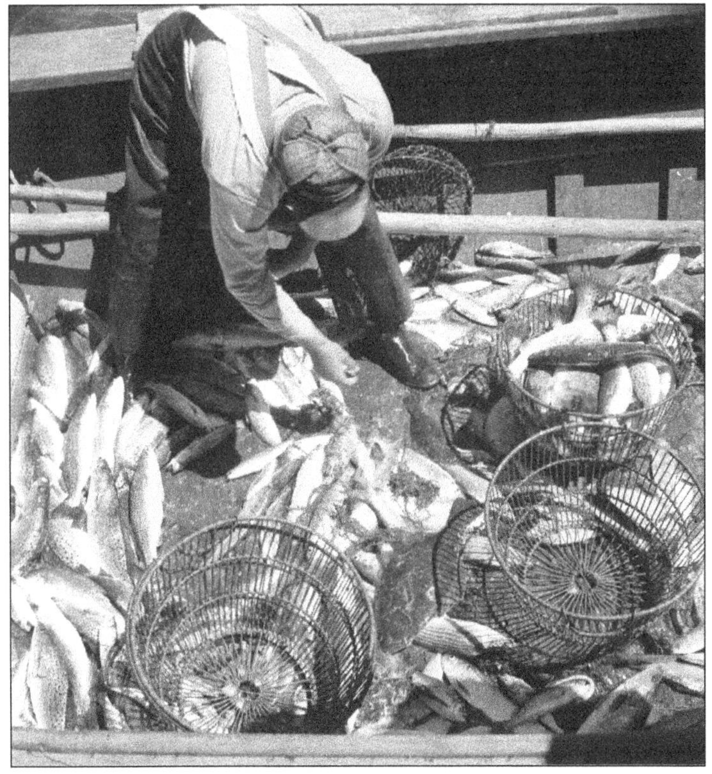

In the same era as this photograph, an addition to the Methodist church was undertaken. Women from the church went to the docks and asked for a donation of a few fish from each boat that came in. They persisted until they had made up a 100-pound box that was included in the next shipment up North. They used the proceeds from that sale for Bethany. (Courtesy of Aycock Brown Collection, OBHC)

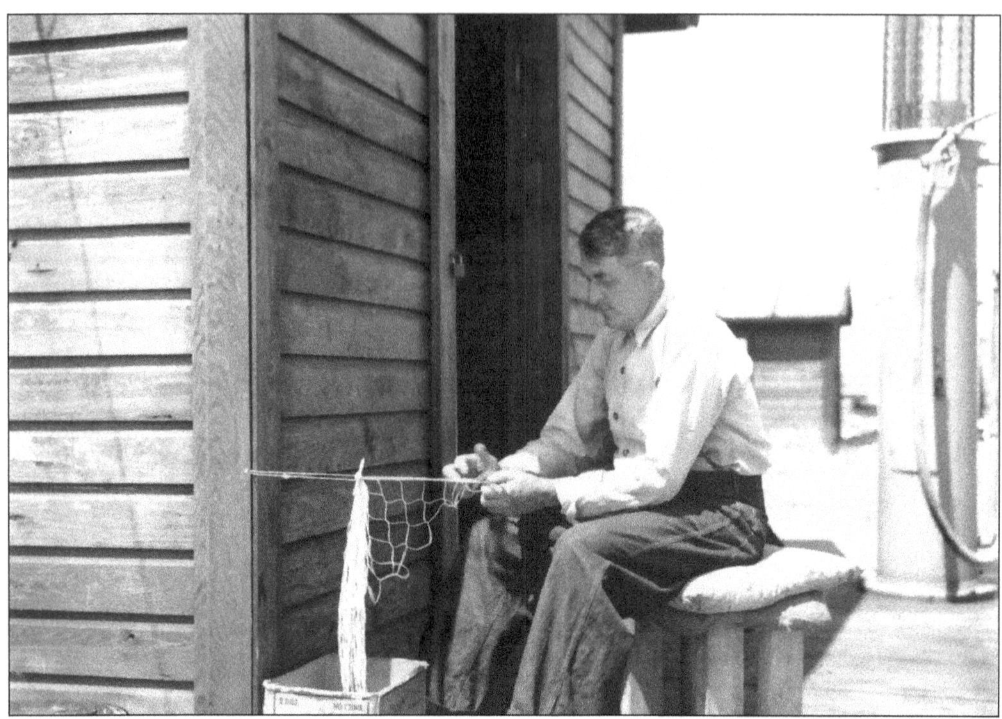

In government-issued work garb, this Works Progress Administration (WPA) enrollee learned the art of net mending. During the New Deal era of the early 1930s, hundreds of men came to Dare County through the WPA and the Civilian Conservation Corps (CCC). They lived in camps on the beach and some were charged with constructing sand dunes. (Courtesy of NPS, CHNS.)

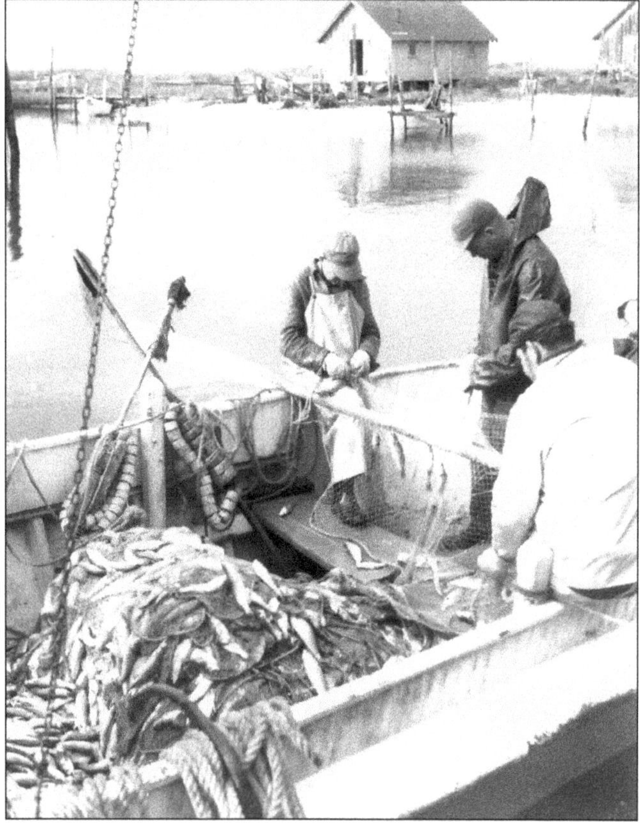

For generations, men in Hatteras had little choice but to fish for a living. When Life-Saving Service and lighthouse-keeping positions became available in the mid-1800s, men competed for those prized federal spots. After World War II, tourist industry-related jobs and ferry service careers gave local men more alternatives for making a living. (Courtesy of NPS; CHNS.)

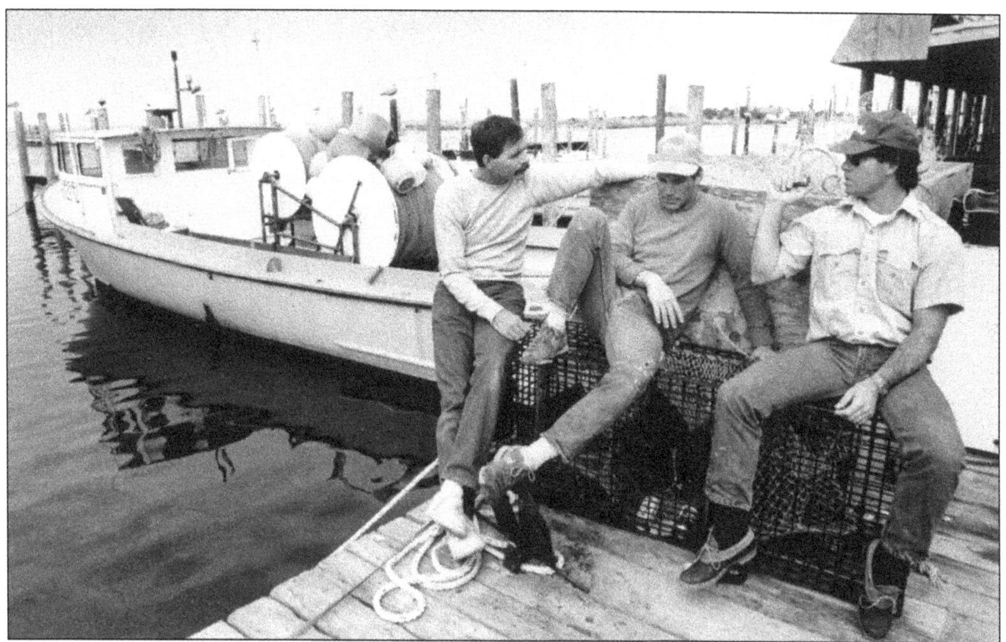

Hatteras commercial fishermen, from left to right, Roger Huffman, Jeff Oden, and Buddy Hooper commiserate about being unable to take out their boats in March 1988. An overabundance of trout brought about a saturation of the market, causing local fish houses to close up shop until the glut dissipated. These men must have felt they could not win for losing. (Courtesy of Drew Wilson Collection, OBHC.)

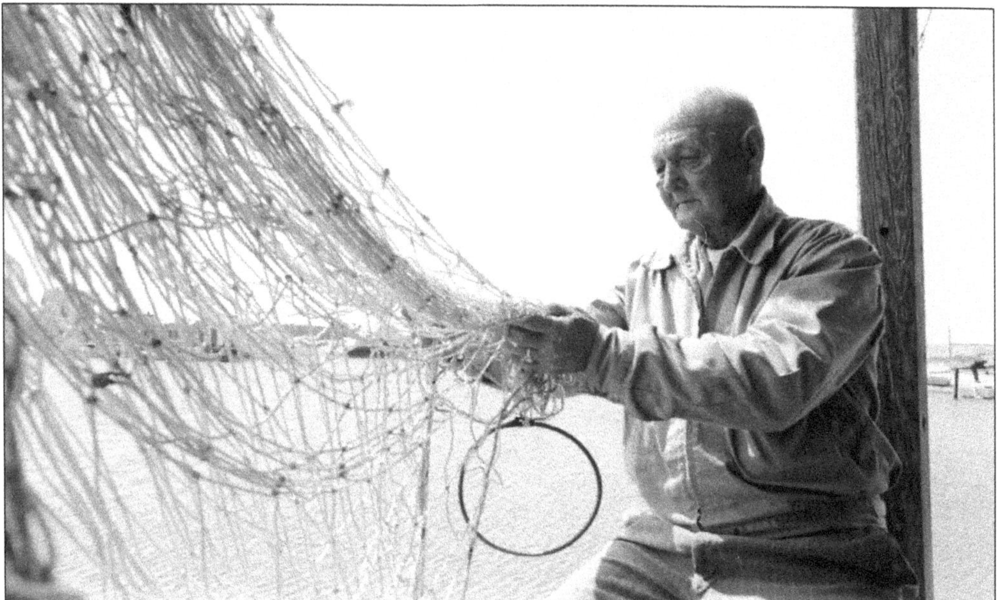

Fishing always comes down to a man and his net. The large meshed gill net was used to catch shad. The iron rings, used before lead weights, were heavy enough to keep the net in place, but light enough to bob up with the tide. Indulgent grandmothers were known to bend a hook in a clothes hanger and let little boys roll the large rings along the ground. (Courtesy of Bruce Roberts Collection, OBHC.)

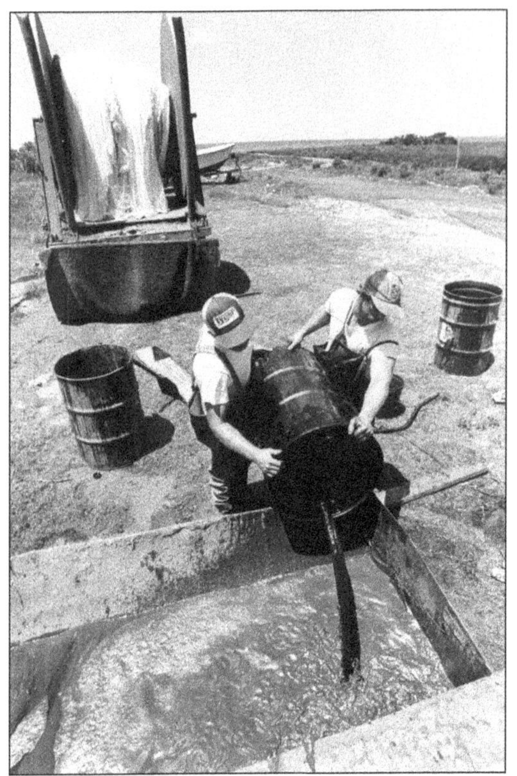

In 1986, Bob McBride and Dave Blackmon of Buxton poured a barrel of tar into a vat in order to treat their nets. Wrapped on a reel in the background, a new bright-white net awaited dunking. Tar is a dark, viscous liquid obtained from the distillation of either wood, coal, petroleum, or peat. The use of monofilament makes that dirty task a chore of the past. (Courtesy of Drew Wilson Collection, OBHC.)

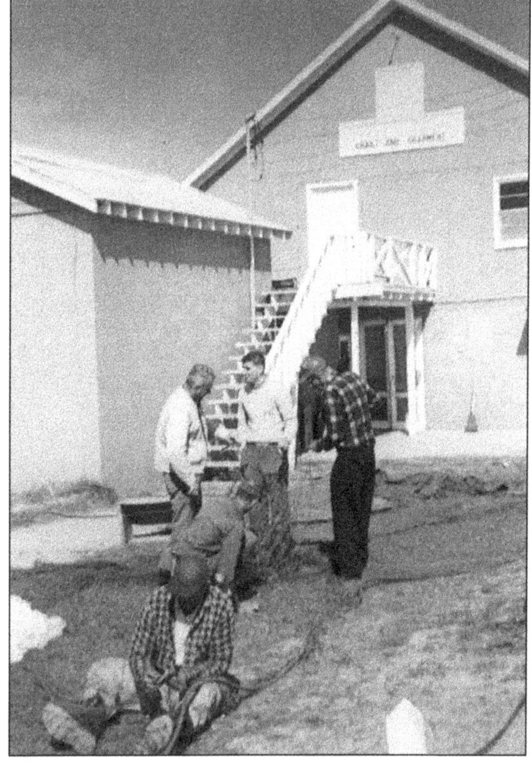

Seashell dealer Woodrow Stetson (on left) was most likely talking to fisherman R.S. Meekins about buying old nets. Stetson sold thousands of decorative kits that contained a length of weathered net and a cork float. Every tourist had to have one. The Stetson family still buys and cuts menhaden nets, realizing that, one day, old cotton or nylon nets will be unavailable. (Courtesy of Robin Holt Daniels.)

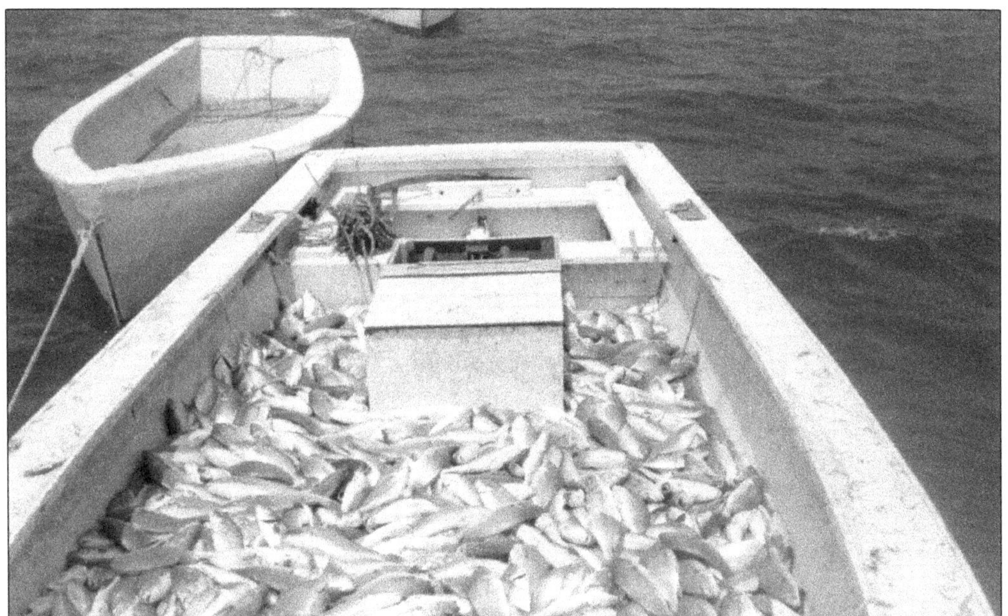

Named for the sound they make, croakers were often caught by the last long netter, Charles Daniels. Now almost deaf, he still remembers hearing his three greatest catches before seeing them. Each time, thousands of croakers roared from his net below the water. These croakers in the bottom of his run boat in the late 1970s, weighed three to four pounds apiece, more than four times what they usually weigh. (Courtesy of Jeffrey W. Midgett.)

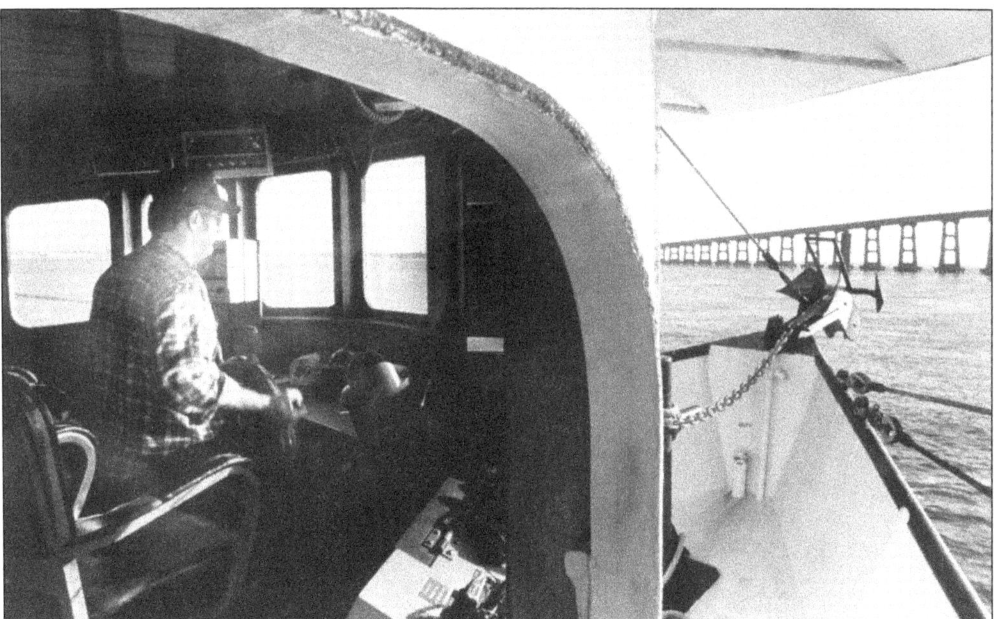

Three days after Christmas 1987, James Craddock piloted the *Capt. Ralph* toward the Herbert C. Bonner Bridge. The Manns Harbor native has spent a lifetime aboard his steel-hulled trawler that he purchased in 1979, seeking whatever species the market accepted. The vessel is very much a part of the family. Craddock's cousins lovingly stitched its image into a quilt for him. (Courtesy of Drew Wilson Collection, OBHC.)

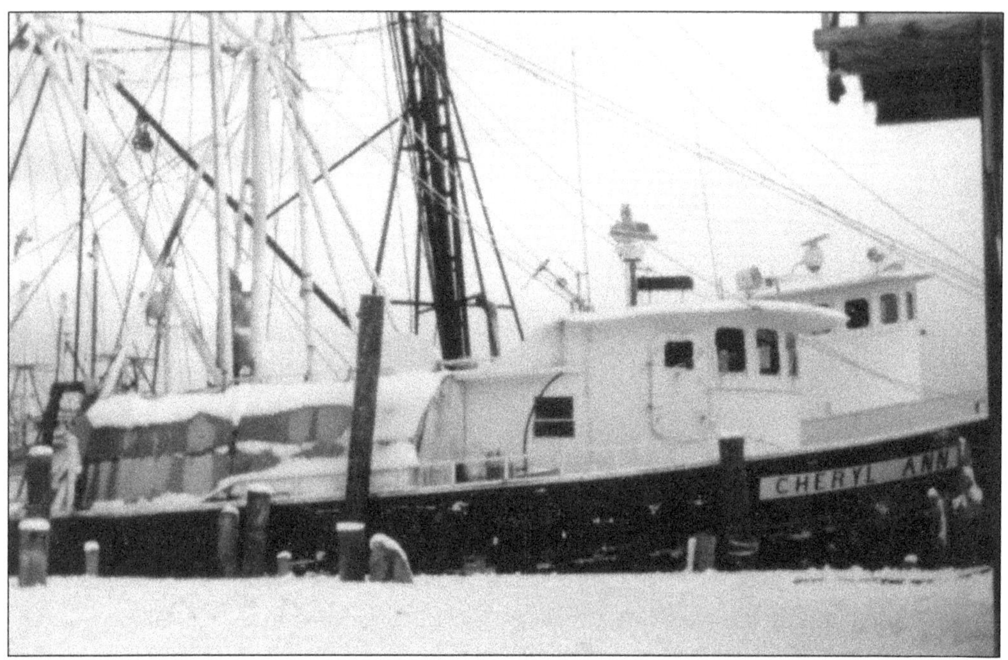

In January 1980, the *Cheryl Ann* was held fast in the ice in Cape May, New Jersey. Capt. "Frog" Tillett took her far and wide in search of flounder, sea bass, and porgy. He had one of the first vanity license plates on the island, which was stamped PORGY, claiming bragging rights for dominating landings of that species. (Courtesy of Frog Tillett.)

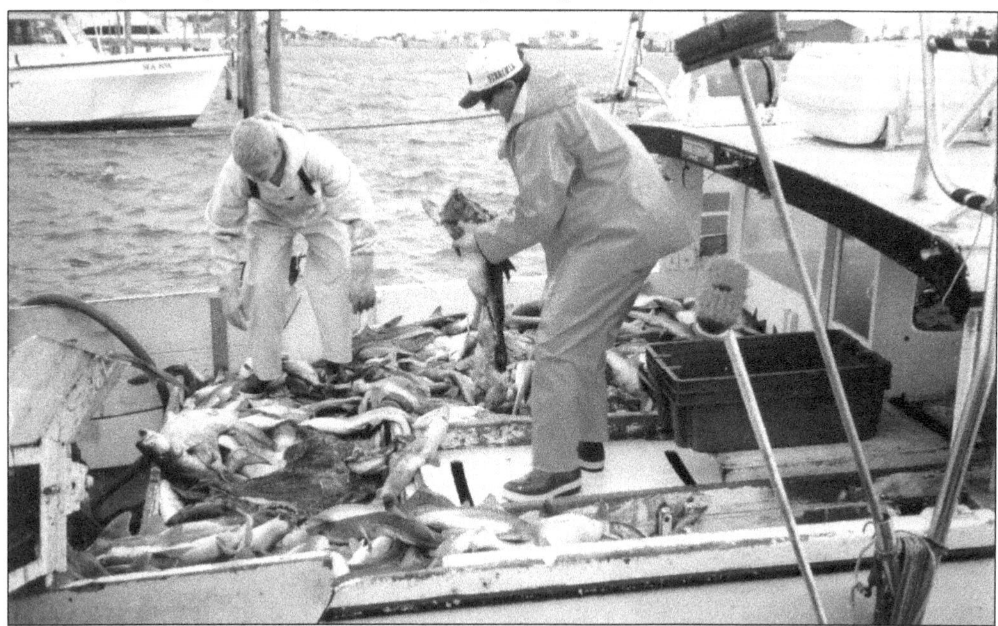

Britton Shackelford (right) found a lone monkfish in a load of dogfish. Previously underutilized, both species were profitable for fishermen as more traditional catches dwindled or were regulated. Monkfish were prized for their lobster-like taste, and dogfish were the staple of fish-and-chip shops in England. Fishermen have always had to diversify their business strategy. (Courtesy of Britton Shackelford.)

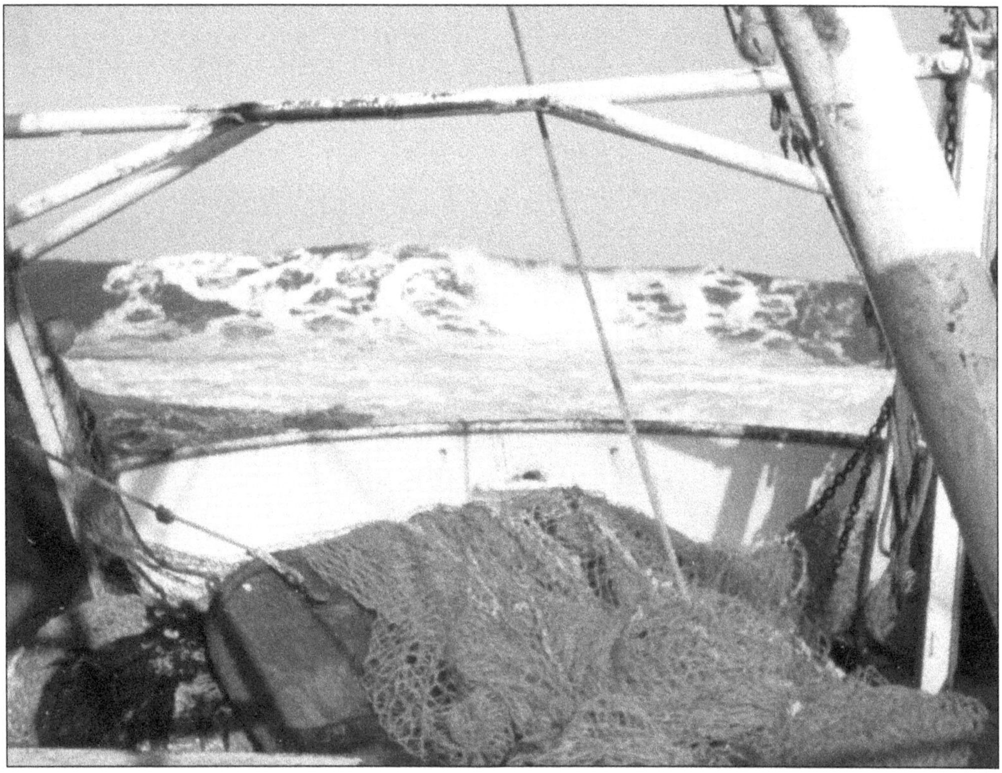

This wave breaking in Oregon Inlet looked as enormous as one of Hawaii's big kahunas to the men aboard this trawler in 1980. With swift currents and an ever-changing bottom, there was no way to predict wave patterns. Historian David Stick said that the inlet is actually an outlet, the only one in a 100-mile stretch that also bears unbelievable water pressure on either side. (Courtesy of Jeffrey W. Midgett.)

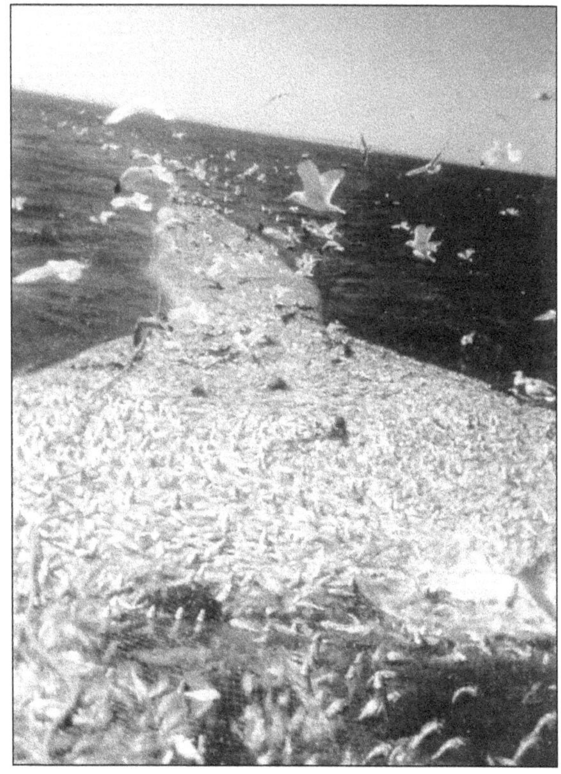

Moon Tillett bought the *Linda Gayle* for Billy Carl for $120,000 because he thought his son needed the safety of the steel hull. For his second son, Craig, the *Gallant Fox* was purchased for $275,000 in 1979. Both Alabama boats were wise investments, paying off with tremendous catches like this one that the seagulls are animated about. (Courtesy of Billy Carl Tillett, photograph by J. Foster Scott.)

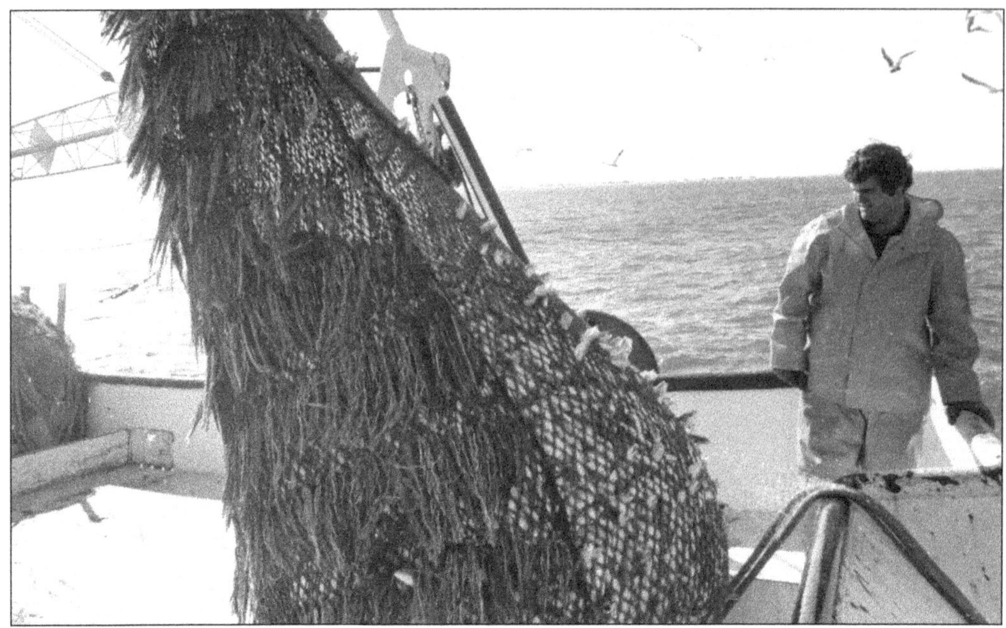

Billy Carl Tillett carried many guests out on the *Linda Gayle*. Scientists, politicians, and authors have all hung tough with the hard-driving fisherman. Photographer J. Foster Scott fared the worst; he was seasick from the moment he stepped off the dock until he returned. Scott earned the crew's respect by never complaining about the rough, cold trip. (Courtesy of Billy Carl Tillett, photograph by J. Foster Scott.)

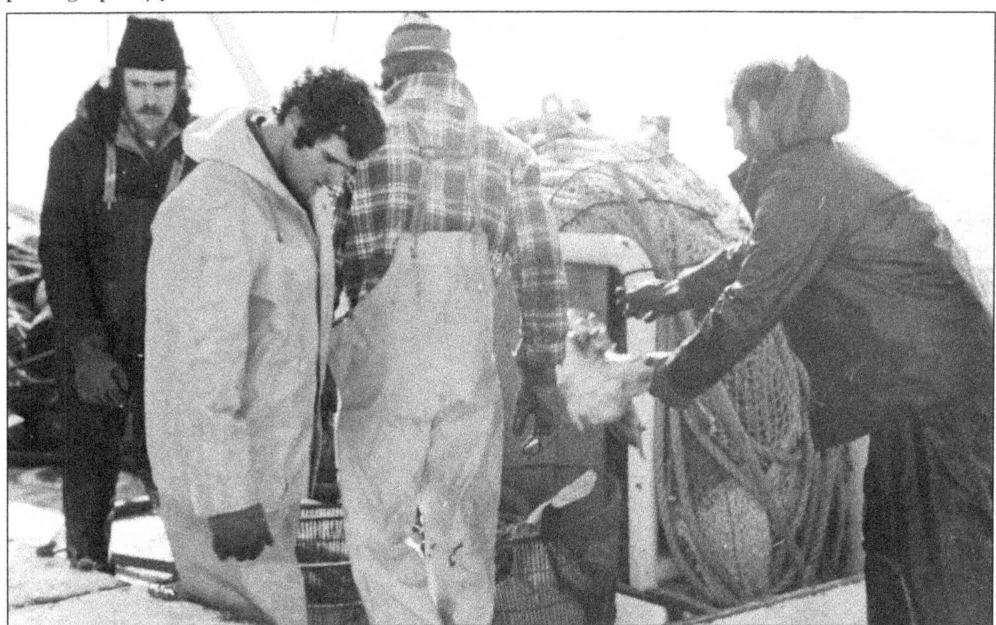

Never one to confine himself to the wheelhouse, Tillett found fishing an exciting occupation, especially when a full net bag floated to the surface. Of today's restrictions to fish within specific parameters, Tillett said, "It's hard to avoid a species when you don't even know what is out there. Quotas take the interest and fun out of it." (Courtesy of Billy Carl Tillett, photograph by J. Foster Scott.)

Not everyone can mend net, and it is unusual that a captain undertook the task. Since meshes (called "marshes" by locals) are squares, it is a simple matter of geometry, but it is a hard problem to decipher. In crews of old, one man was assigned to mending the net, and he was constantly busy since the net was torn almost every time it was used. (Courtesy of Billy Carl Tillett, photograph by J. Foster Scott.)

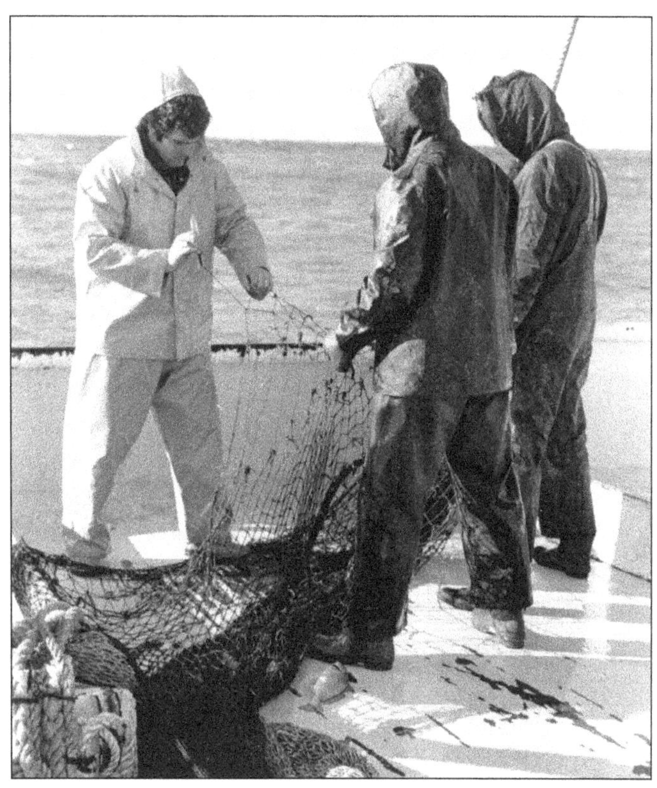

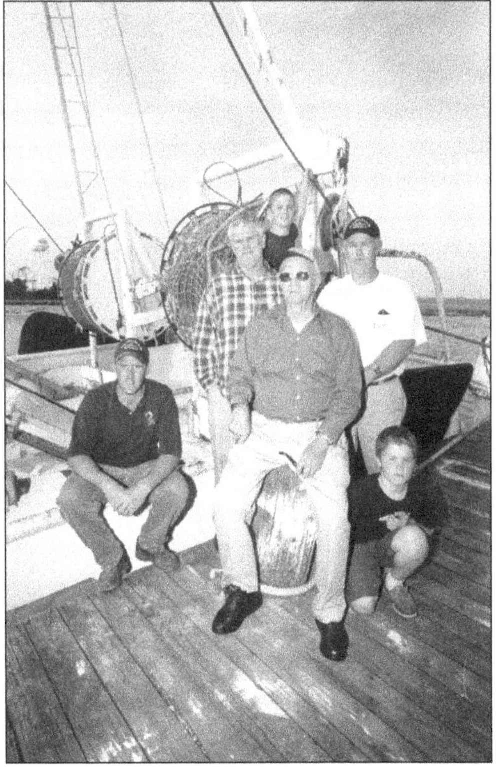

Tillett men, from left to right, Ryan, Billy Carl, Brock, Moon, Craig, and Ramey posed on the Moon Tillett Fish Company dock in the early 2000s. Founded in 1977, the business greatly expanded when it embraced volume and shrewdly played the market by freezing and holding seafood until the price went up. In addition, the Tilletts initiated trade with international brokers. (Courtesy of Billy Carl Tillett.)

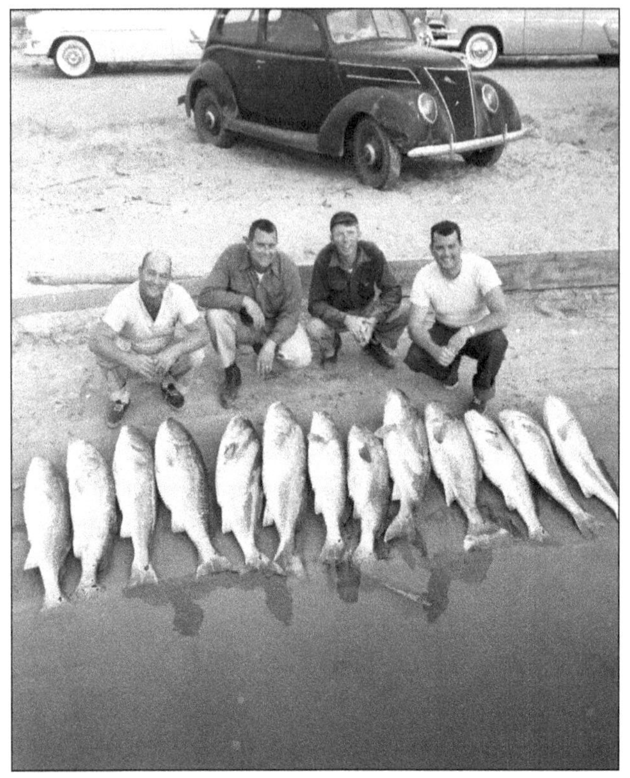

Watermen can be hardheaded and opinionated, but there is one thing they all agree on: Willie Etheridge Jr. (third from left) was the best ever. He set standards in both the charter and trawl boat industries. From 1963 to 1983, he fished for lobster, flounder, and shrimp from Massachusetts to Florida, constantly trying new gear and sharing knowledge with others. (Courtesy of Aycock Brown Collection; OBHC.)

Willie Etheridge Jr. would steam quickly back to the fish house with the first part of his catch, securing the highest price before the market broke. Then he would go back and gather the rest of his haul, knowing that it was better to be paid top price for a small percentage of his fish than a lower price later for all of them. (Courtesy of Billy Carl Tillett, photograph by J. Foster Scott.)

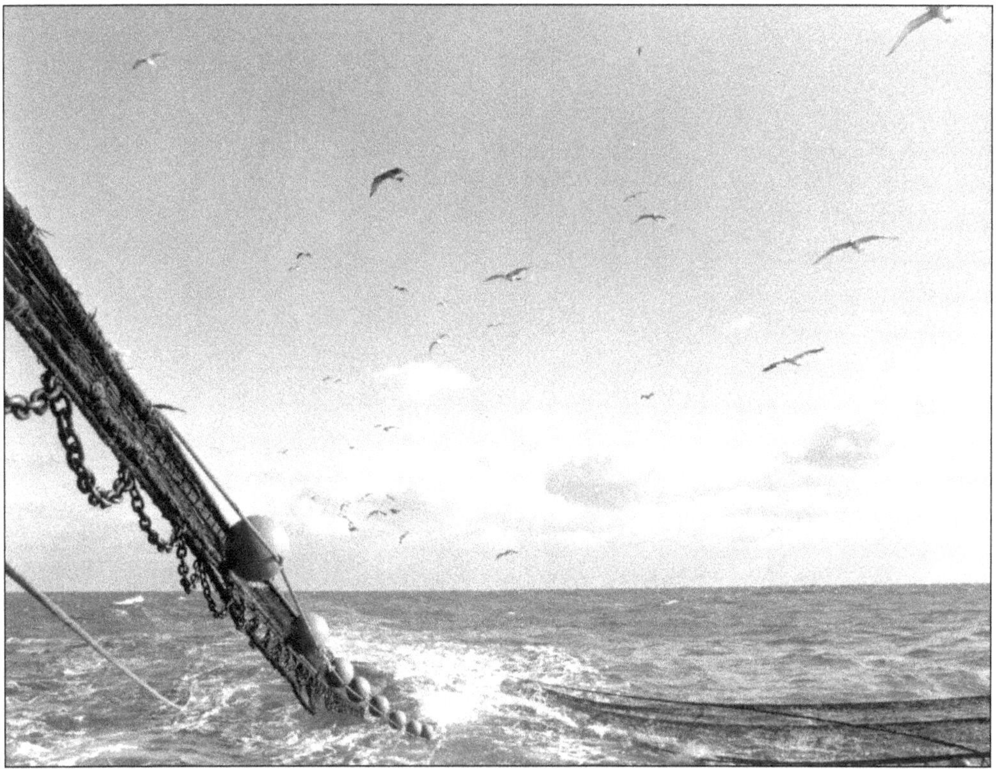

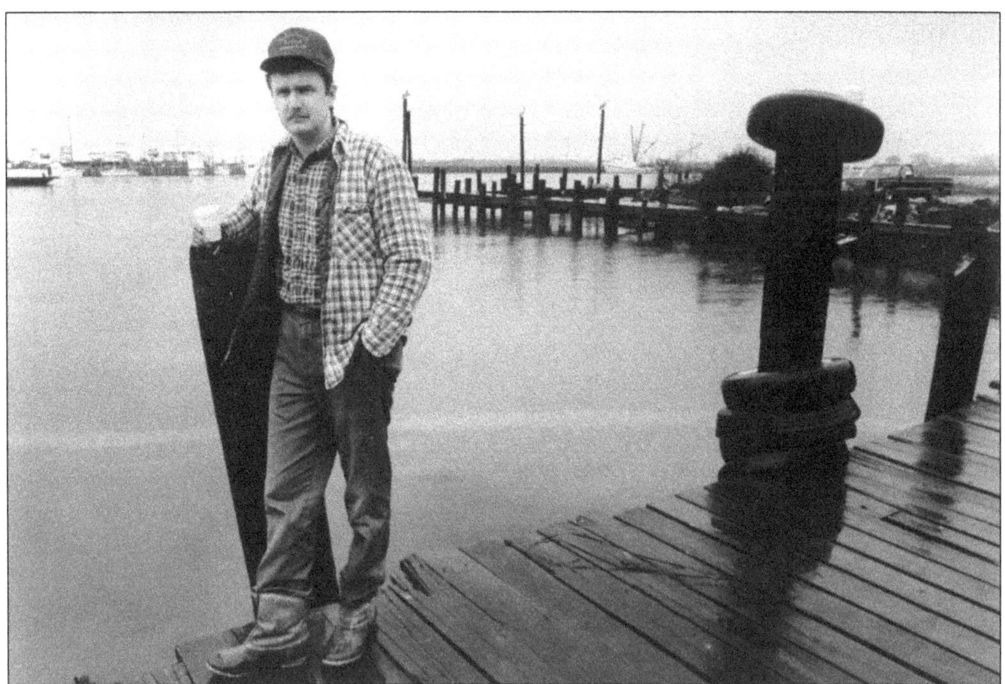

On March 24, 1989, assistant manager Rex Simpson had his coffee on the dock of Fisherman's Seafood, where no boats had moored in three weeks, due to shoaling conditions at Oregon Inlet. Simpson worked in seafood sales and ran his own restaurant and seafood trucking firm until his uncle made him an offer he could not refuse—buying his roofing company. (Courtesy of Drew Wilson Collection, OBHC.)

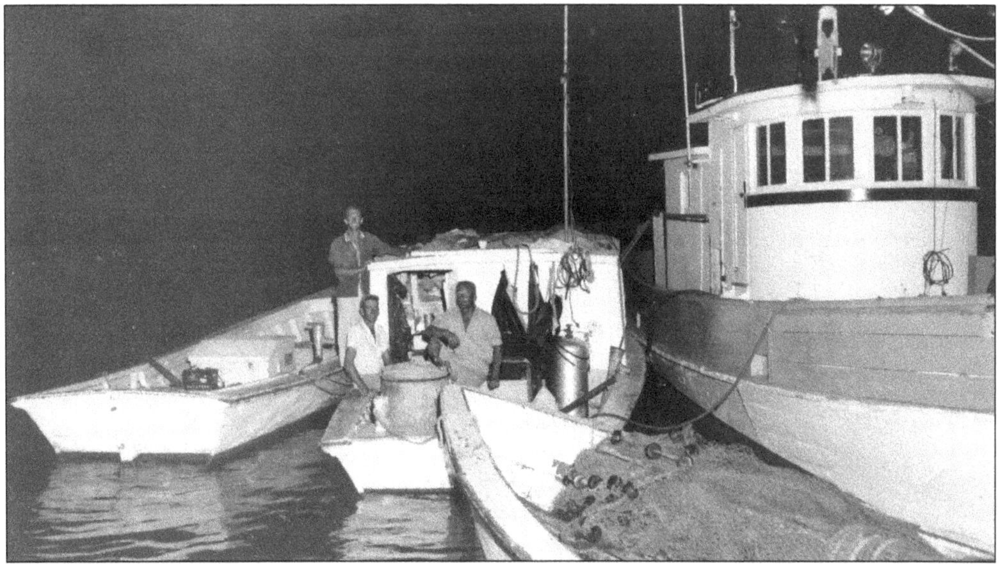

Not-so-distantly related, Bob, Dennie, and Charles Daniels were long netting in Hatteras when their picture was taken. The newspaperman from Bellerose, New York, was interested in the friendly crow that had taken up with the fishermen for the season. He was perched on top of one of the cabins but did not show up against the night sky. (Courtesy of Robin Daniels Holt, photograph by Thomas Begley.)

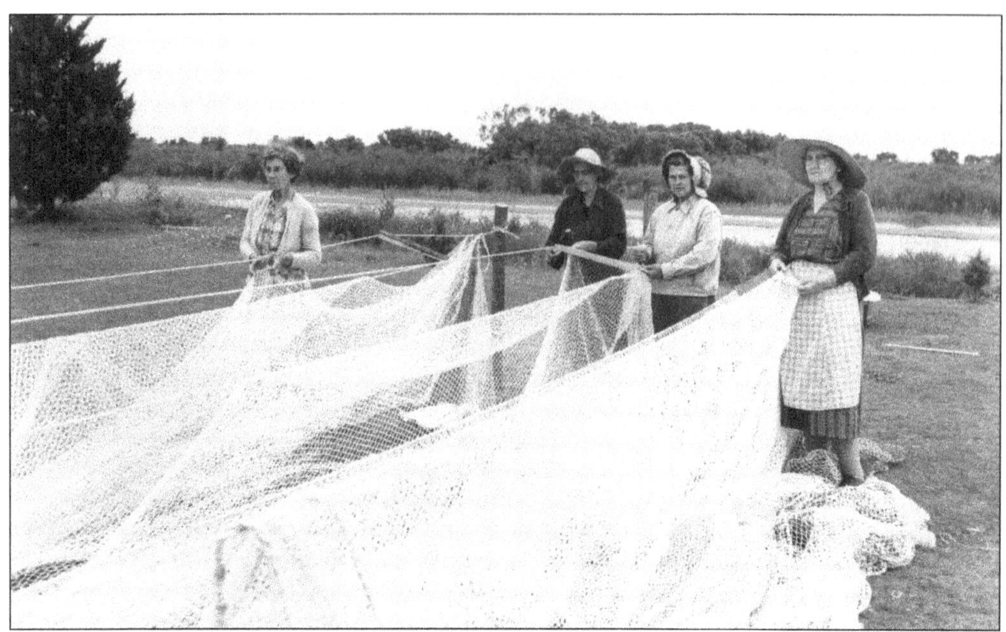

In each sound-side community, women helped the men by hanging yards of cotton netting. On a cool May day in 1954, Stumpy Point ladies stretch out new webbing in Dick Best's yard before hanging it on the rope with a series of half-hitch knots using large weaving shuttles called net needles. The practice of women hanging net died out with this generation. (Courtesy of Roger Meekins Collection; OBHC.)

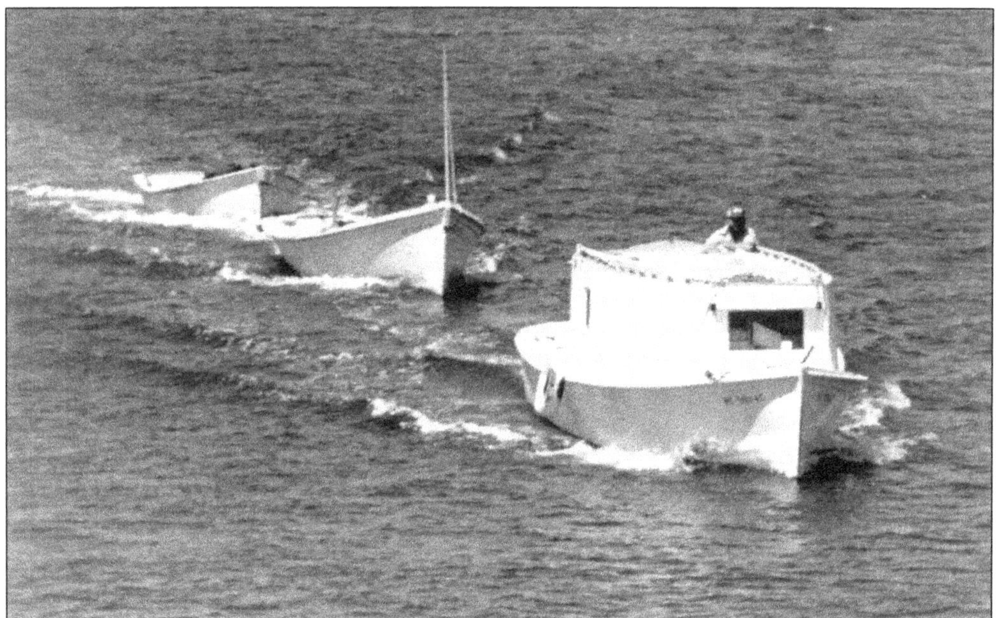

Elisha Meekins stands on the washboard of the lead boat of his father's long net rig. At age 93, his mother, Joyce, can still remember a whopping catch in June 1981 she helped bring in. They had to shuttle boatload after boatload from the rig to the dock, and the "run boat" was riding so low that she had to work the hand pump all day and night on every trip. (Courtesy of Joyce Meekins, photograph by the *Coastland Times.*)

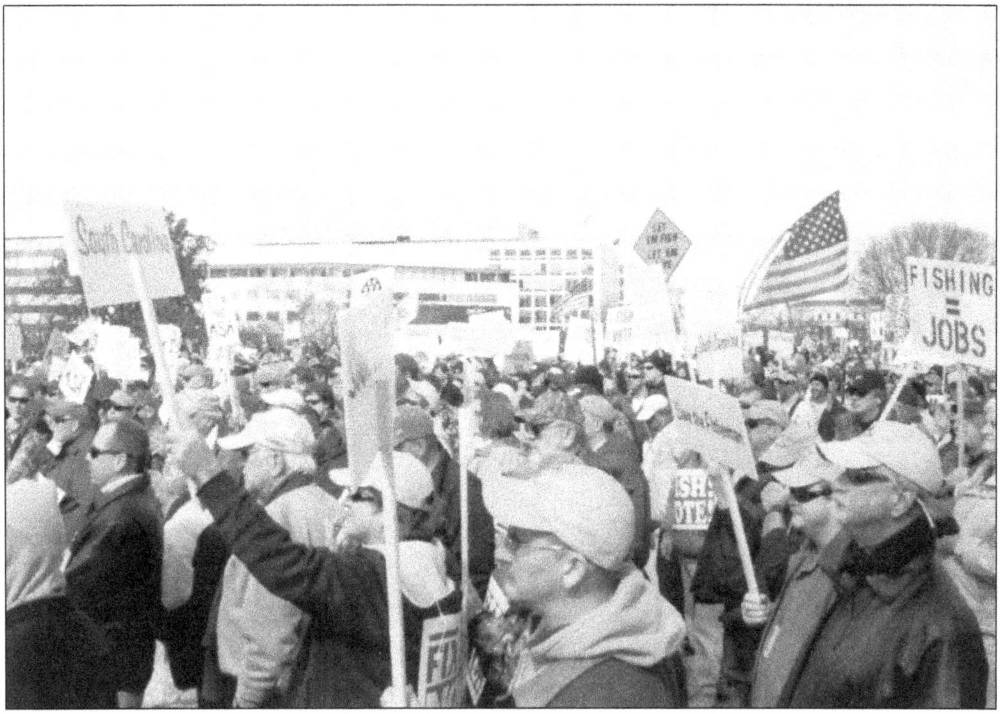

The fiercely independent spirit that most commercial fishermen embody has proven detrimental in their efforts to organize as a group. There have been attempts to assemble like this Fishing for Freedom rally in Washington, DC, but, by and large, they have not been able to play the political power game. Their collective voice has only been a whisper to policy makers. (Courtesy of Britton Shackelford.)

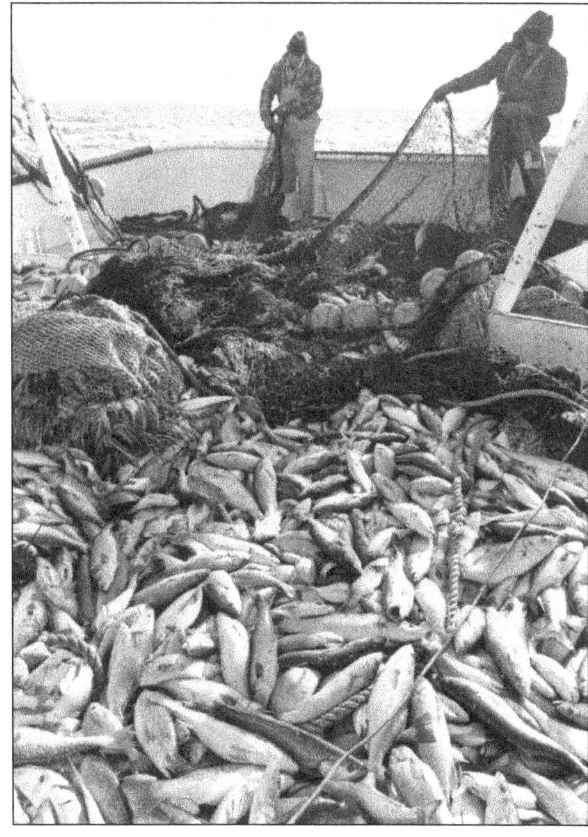

According to author Robert Fritchery in *Wetland Riders,* the building animosity toward commercial fishermen began in Louisiana and Texas when oil industry businessmen, who fished recreationally, used their deep pockets to lobby state politicians. From their lobbying efforts, a perception evolved that commercial fishing is not compatible with good stewardship. (Courtesy of Billy Carl Tillett, photograph by J. Foster Scott.)

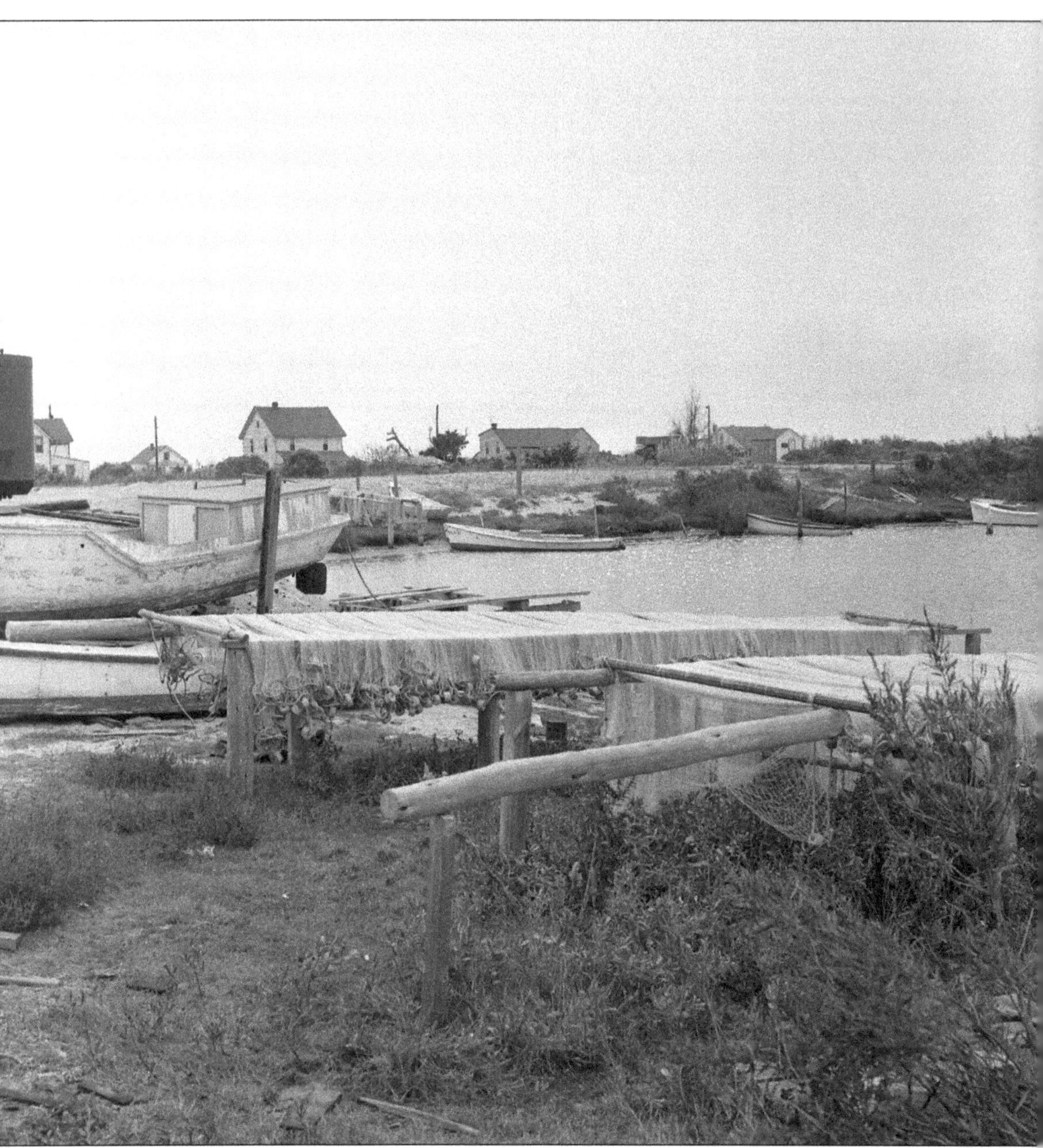

In 1993, author Robert Fritchey wrote, "Like a Zydeco accordionist, the fisherman learns his trade from another. There are no books or schools that teach you how to catch enough fish to live. When the fishermen are gone, they are gone forever. Unfortunately, their value extends even beyond the food that they provide, the knowledge that they hold, and the colorful tradition they add to coastal cultures. As those who earn their livelihoods directly from the coastal marshes are eliminated, so too is their strong interest in protecting this threatened habitat." His words that may have seemed somewhat radical a quarter of a century ago, are more disturbing now as the move to phase out the commercial fisherman has become a very real objective. (Courtesy of Aycock Brown Collection, OBHC.)

Six

LOCAL MAINSTAYS

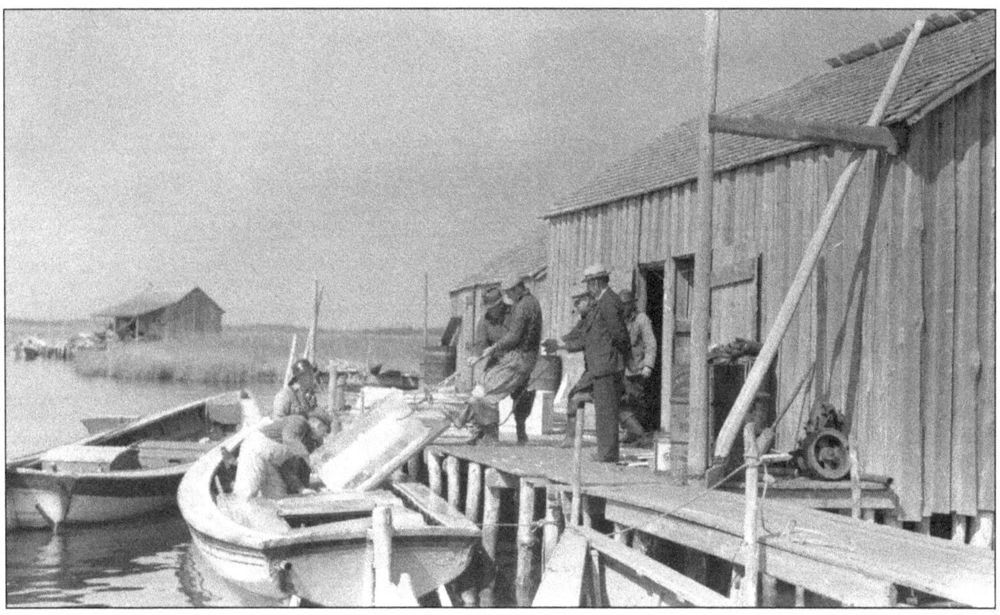

On the east side of Roanoke Island, Mill Landing Harbor, pictured here, was originally about 10 inches deep. It was dug out in 1933, and in 1946, a dredge more expertly deepened the harbor, casting spoil up onto the land. Globe Fish Company's depot was located on the west side of the village at the Old Wharf, where the water was historically much deeper. (Courtesy of SANC, photograph by Charles Farrell.)

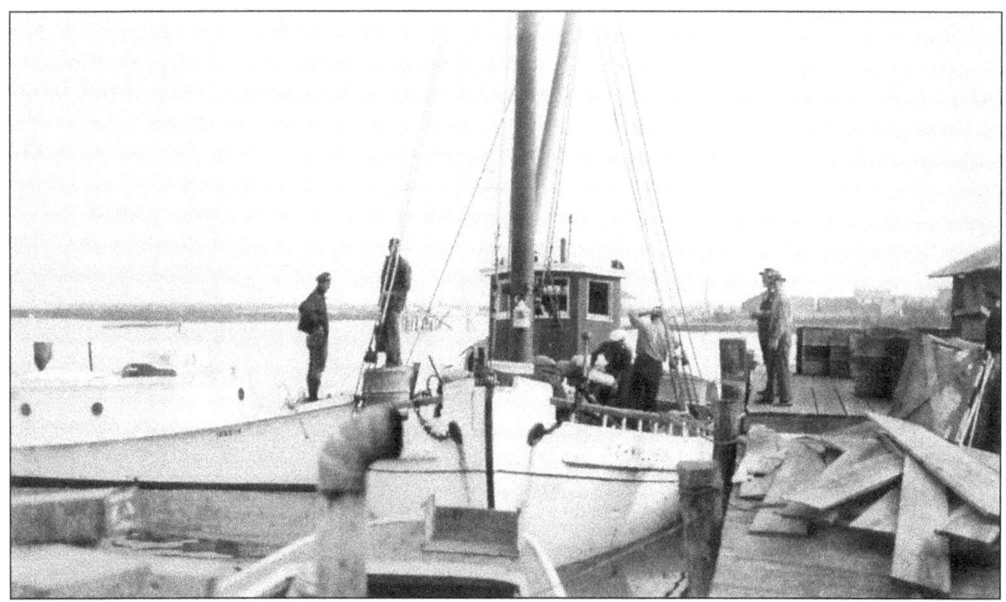

Part of the "Wanchese Line," the *CH Mallison* transported tons of fish from Globe Fish Company stations to Elizabeth City. The Outer Banks has shallow draft harbors, with dredging occurring since before the Civil War. The propeller action of the *CH Mallison* dug out a 20-foot underwater hole at the Old Wharf. Many boys learned to dive off the cabin of the docked freighter. (Courtesy of Museum of the Albemarle.)

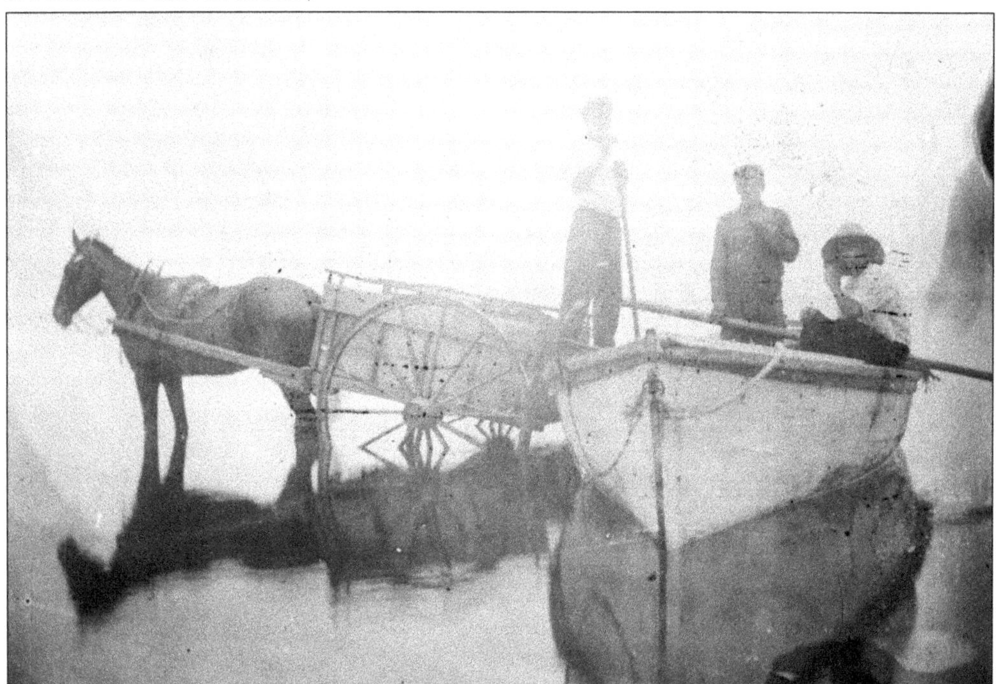

These fishermen used shallow water and a slick-calm day to their advantage by backing a horse and cart right up to their shad boat. Modern-day pound netter Lee Craddock said that he loved the work even though it was hard. It takes a lifetime of observation and much understanding of the habits of fish to set a trap in the right place. (Courtesy of D. Victor Meekins Collection, OBHC.)

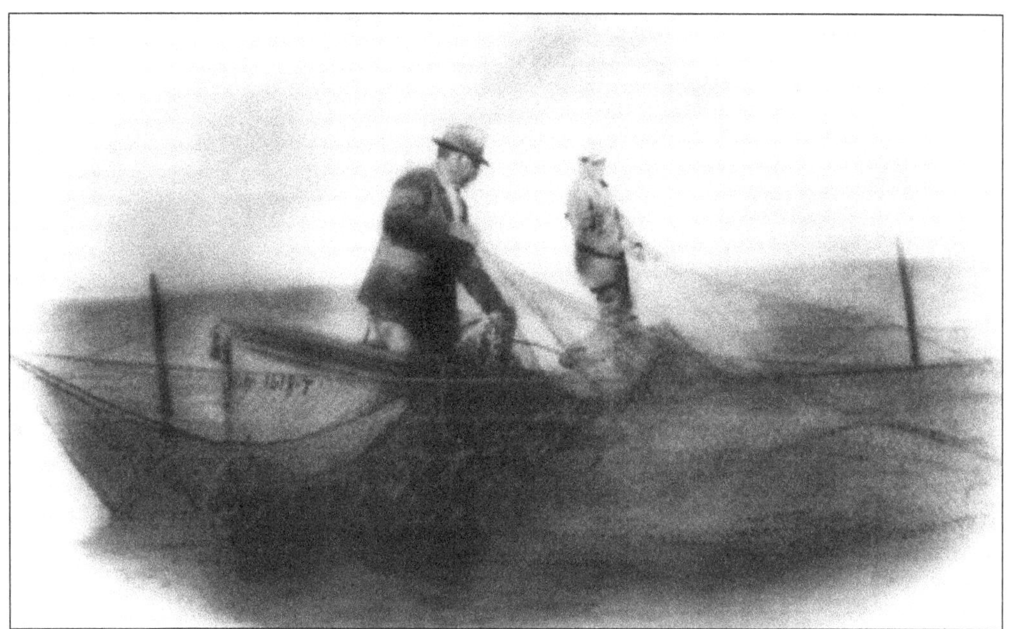

The Shannon men were pound netters in Manteo's Shallowbag Bay. The canal behind Darrell's Restaurant is still known as Shannon Ditch. Fishermen would have to go into the woods to cut the 20-to-25-foot gum tree stakes that would then be beaten or jetted down with water pressure into the floor of the sound. Fishing legend Wayland Baum fished pound nets in the ocean with his father. (Courtesy of Miles Creef.)

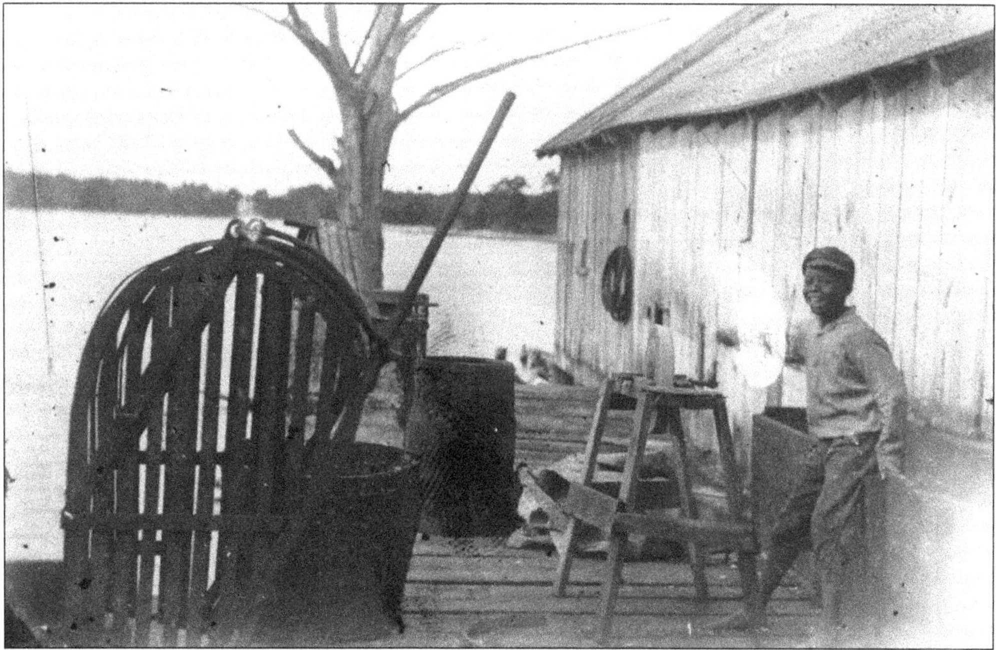

Now a neighborhood, Skyco once had a busy harbor and was the location of various seafood canneries that operated from a wooden building perched on pilings over the water. The large crab-steaming basket and knife-sharpening wheel have caused many to speculate that this happy boy was on the dock by the cannery. (Courtesy of OBHC.)

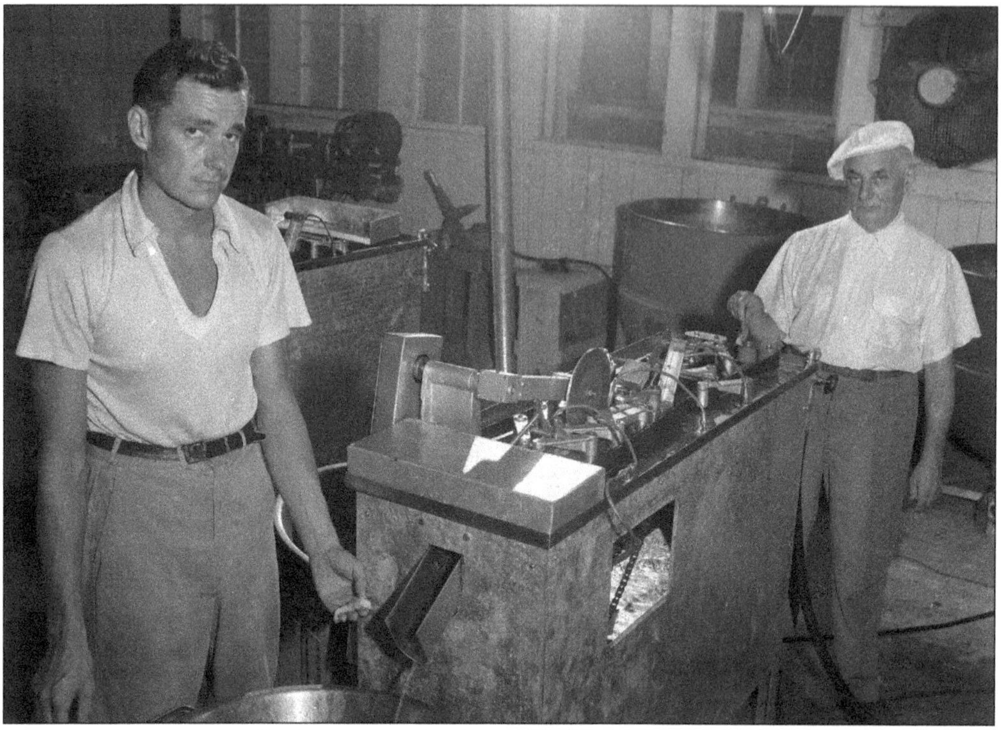

Two stone-faced men coolly demonstrated a shrimp-cleaning machine in Skyco in 1952. Another pair of entrepreneurs, like these out-of-towners, started a shrimp factory by the old Davis Landing in Wanchese. They employed 24 workers who cleaned shrimp to be shipped. It was said that they went out of business after a few years for being big spenders. (Courtesy of Aycock Brown Collection, OBHC.)

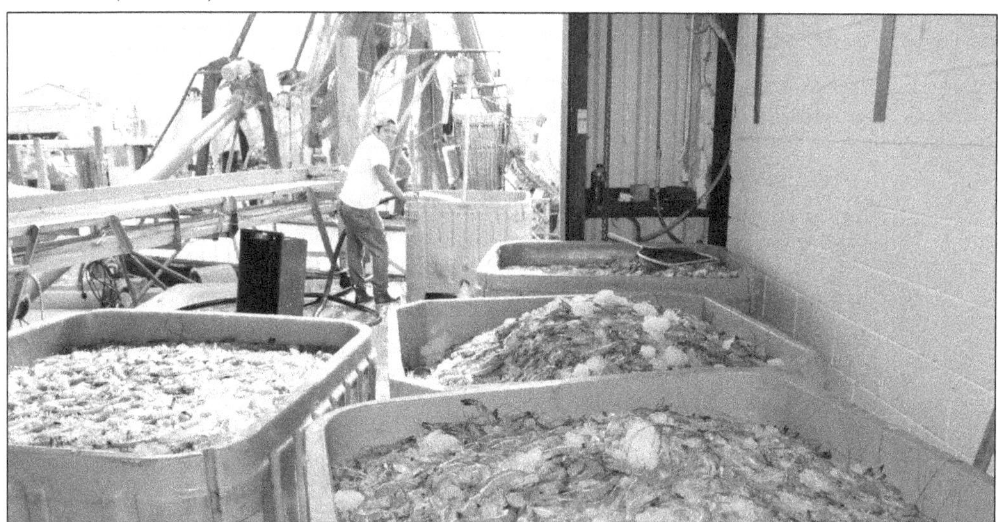

Colonial farmers plowed under unappreciated shrimp to fertilize their soil. Commercial shrimping got a foothold in North Carolina in the 1930s, and by 1939, it was the state's top fishery. It was not until the 1950s that Outer Bankers entered into the game. Until then, shrimp were considered trash seafood with locals calling them "bugs" and refusing to eat them. Clearly, things have changed. (Courtesy of RWG.)

Many men, like Miles Creef, shrimped as a fun side hustle and also as a way to fill family freezers for a year. In February 2018, the North Carolina Marine Fisheries Commission recommended changes in the definition of a licensed commercial fisherman. Seen as an attempt to cut out part-time watermen, the issue faded away, albeit temporarily, after challenged. (Courtesy of Drew Wilson Collection; OBHC.)

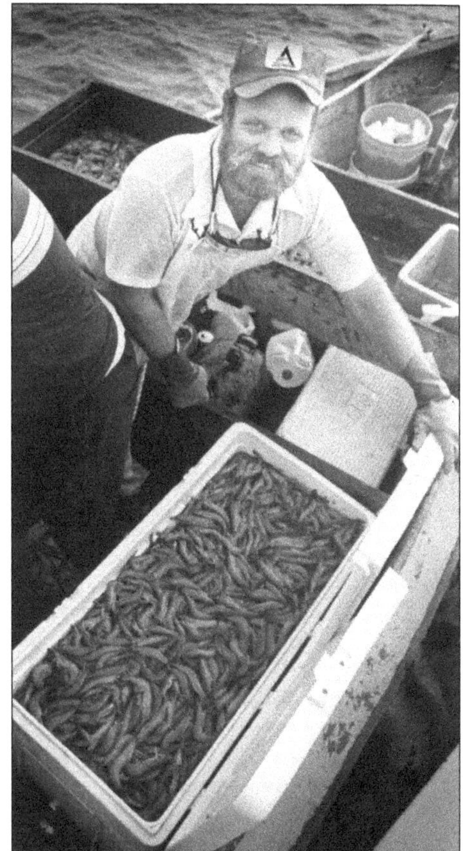

A large vacuum tube is used to suck the shrimp from the hold of the *Capt. Ralph* in November 2017. Shrimping has remained strong in the last years with the season extending into the winter months. Fishermen have been called the "farmers of the sea," but their crop is swimming and that movement changes from year to year. No one knows where or when the shrimp will show up. (Courtesy of RWG.)

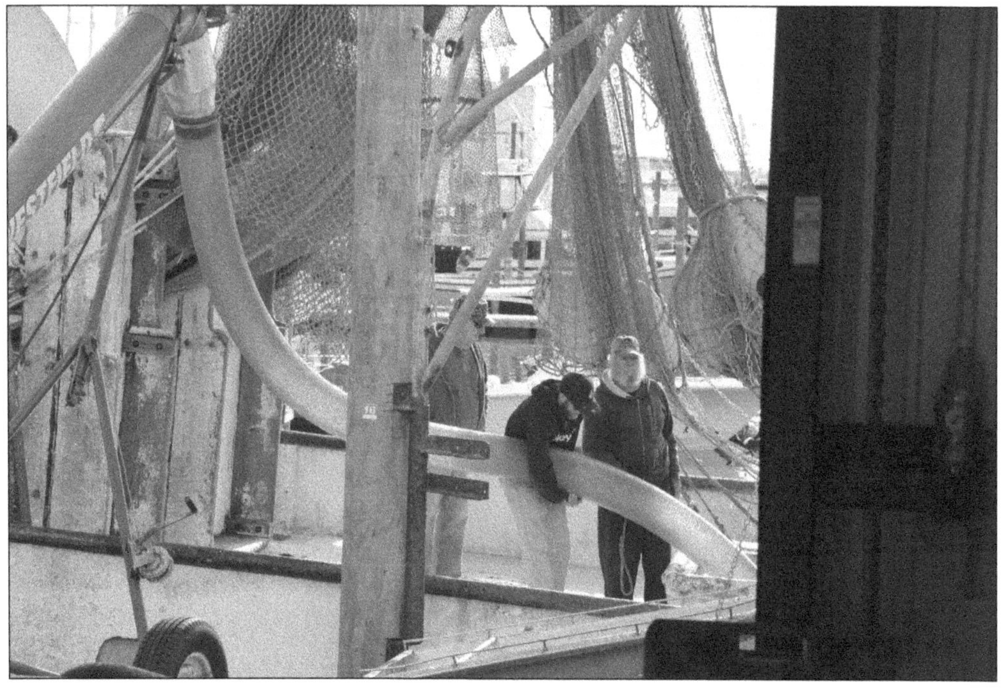

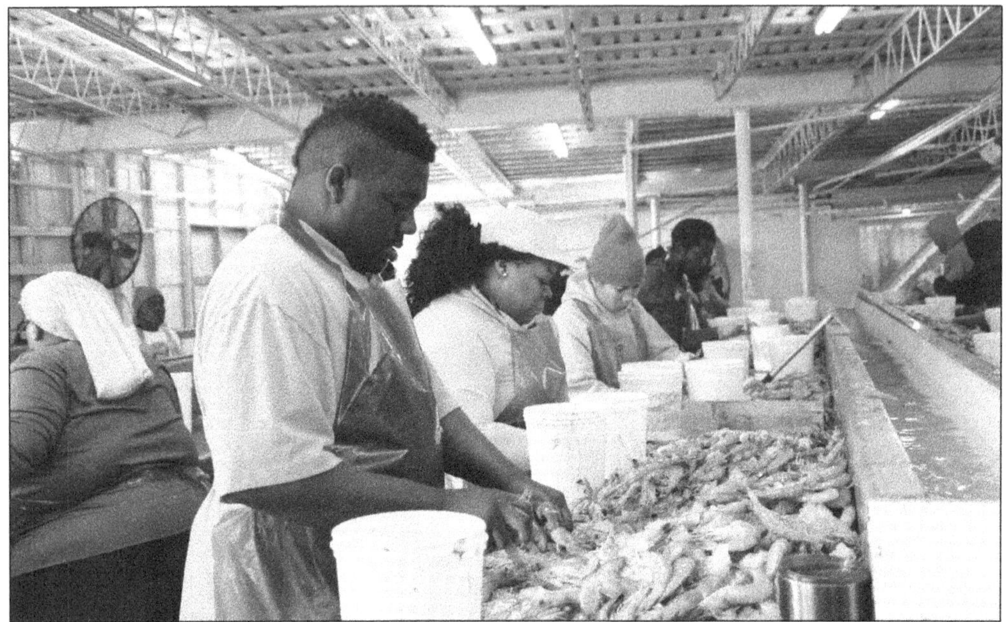

Social media and group texts are two ways that fish houses let potential workers know that shrimp boats are headed to the docks and there is a day of work available. Laborers drive in from neighboring counties for the opportunity to stand in a wet environment heading shrimp all day. Since headers are paid by the pound, a decent, but brief, wage can be made if a person works quickly. (Courtesy of RWG.)

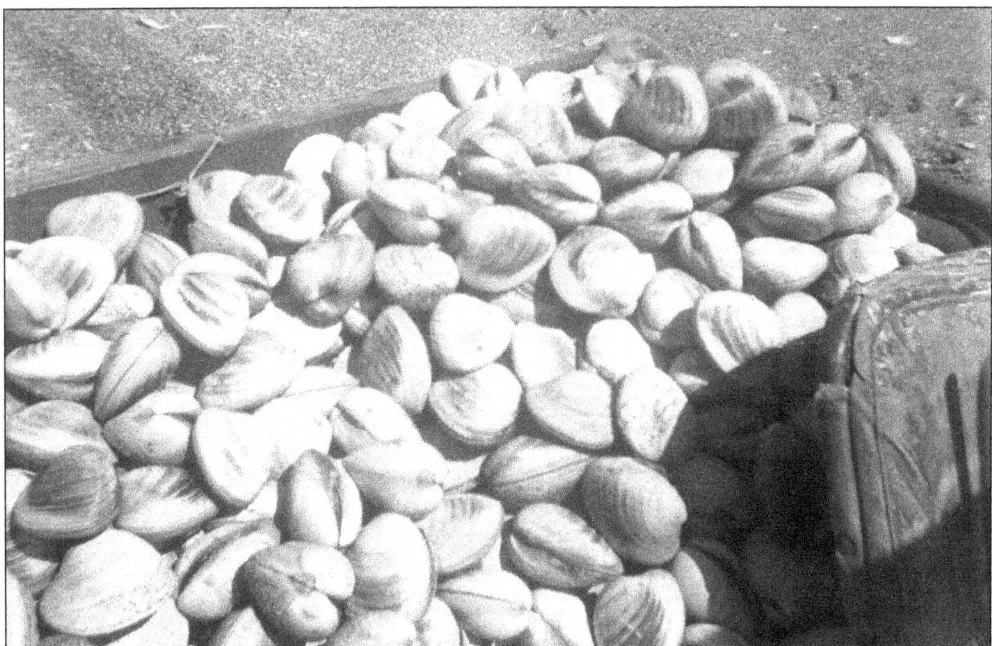

Sea clams as big as softballs once washed up in Corolla. Today, only some local retail markets and restaurants utilize clams in their shells. Commercial clamming by raking or gigging remains a small fishery. It was not always this way. Around 1897–1898, a Long Island, New York, clam factory relocated to Ocracoke Island because of an abundance of quahogs. (Courtesy of Jeffrey G. Midgett.)

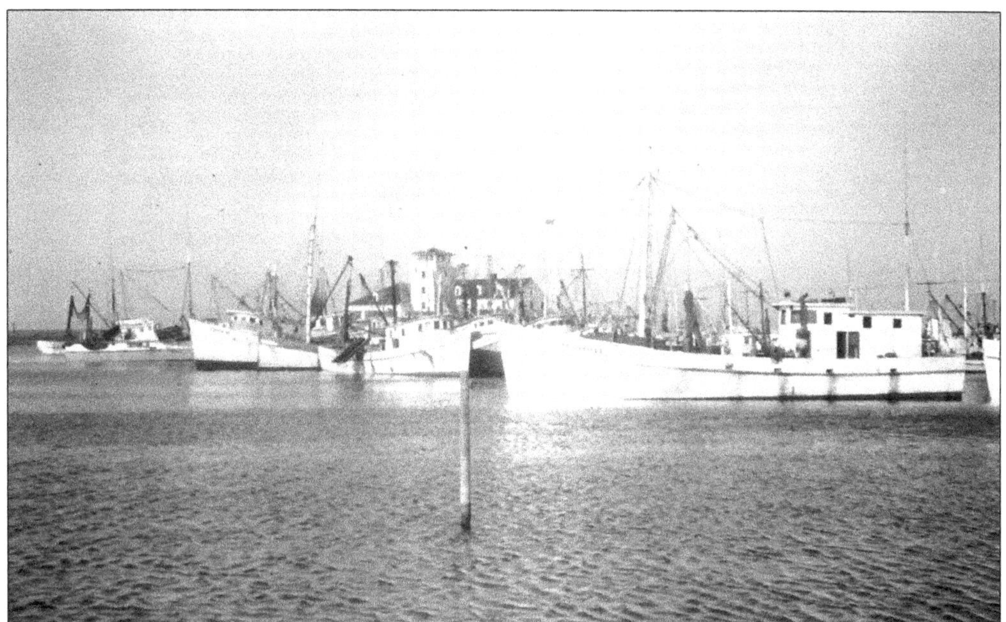

At the turn of the 20th century, Doxsee Clam Processing Facility was built at the entrance to Ocracoke's Silver Lake Harbor, pictured here in a later era. The owner, Henry Doxsee, also built a hunting lodge, boardinghouse, and a two-story residence. The clams bought from local fishermen were steamed in one building and then dumped onto long wooden tables to be opened for canning in another building. (Courtesy of NPS, CHNS.)

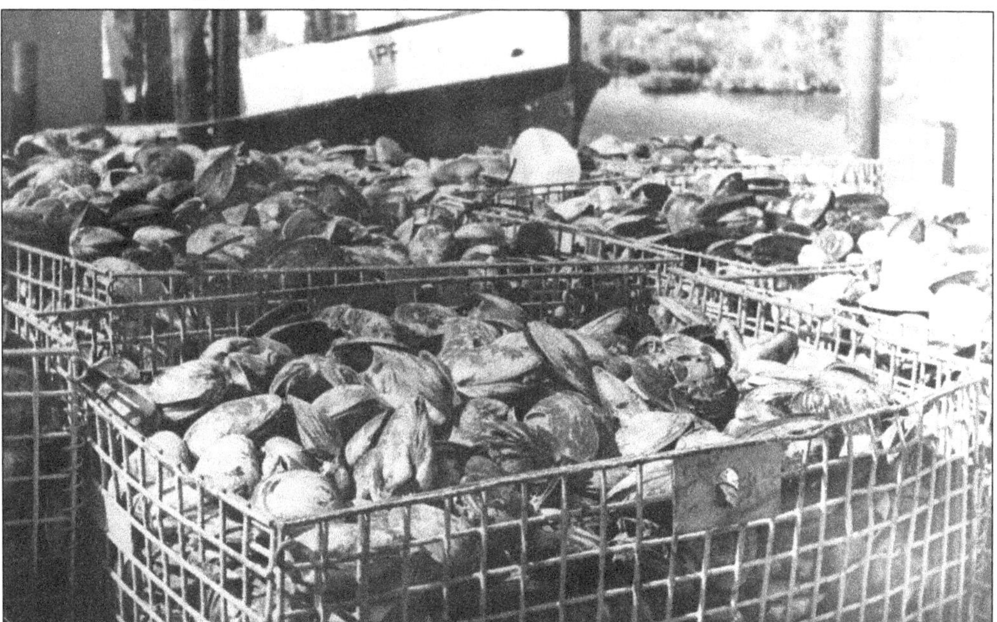

The Doxsee Clam Processing Facility employed young, unmarried women and widows to shell, wash, and pack the clams in their own juices in cans. It was the first time island women could enjoy employment, and they started at 6¢ an hour. The empty shells were tossed out open windows. By about 1910, the clams were overharvested, and the company moved on to Marco Island, Florida. (Courtesy of Aycock Brown Collection, OBHC.)

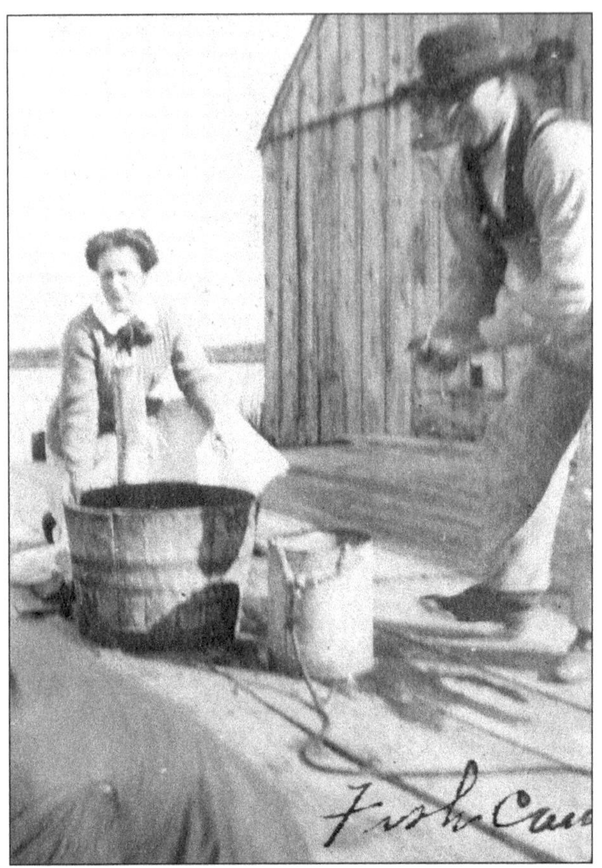

It is not known if Dr. and Mrs. Ropp were visitors or residents of Colington, but it is clear that they were fond of the fish-camp life. If clammers had too many clams to be readily sold, they would take them to a special spot in the water and mark the area with a stake or an enclosure. They figured they had three days to retrieve them before the clams would move away. (Courtesy of Lish Meekins.)

Edward "Jughead" Etheridge, of Manteo, replaced nylon webbing in his oyster dredge. Since oysters attach themselves to each other in a bed, the dredge acts as a rake to bring them up. Naturally occurring oyster beds are rare today. Oysters are missed as a commercial product and also because they improve water quality. One bivalve filters up to 25 gallons a day. (Courtesy of J. Foster Scott Collection, OBHC.)

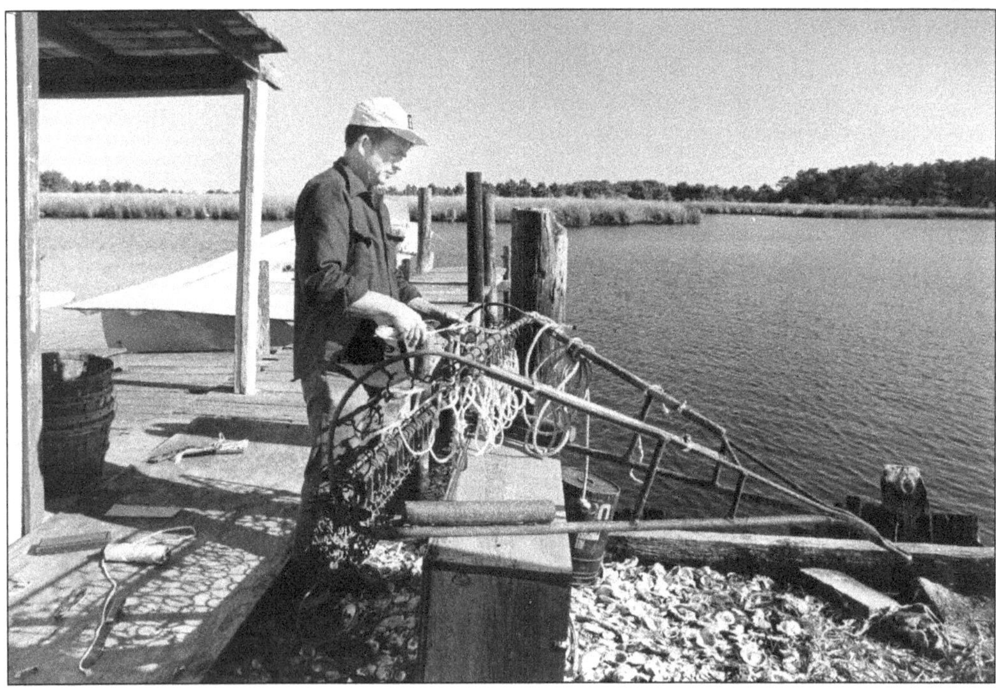

With precarious footholds and wearing street clothes, this crew does not give off the air of a professional oyster operation. An oyster boom in the 1890s caused oystermen who had depleted their shellfish in other states to descend upon Ocracoke. They desecrated rich beds with dredges and longer tongs, violating North Carolina law. Ocracokers gave them a stern warning. (Courtesy of Aycock Brown Collection, OBHC.)

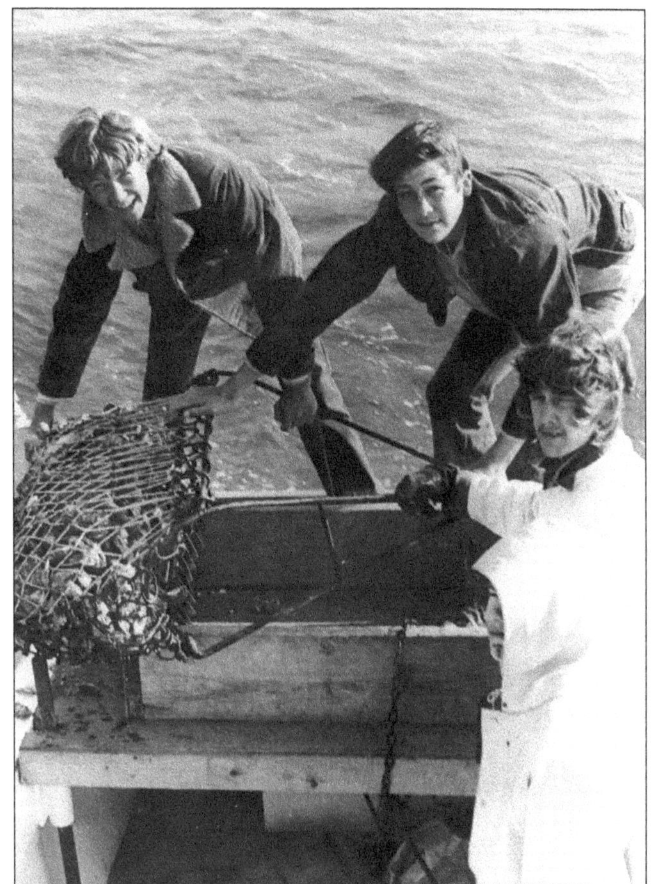

Using a series of pulleys, a bearded oysterman is able to work alone. When the shellfish bed "squatters" continued their plunder, one oysterman found himself at the business end of an Ocracoke shotgun, which he commented had never looked so large. The National Guard arrived on the *Nellie B. Dey* armed with a howitzer and ready to arrest illegal oystermen. (Courtesy of Drew Wilson Collection, OBHC.)

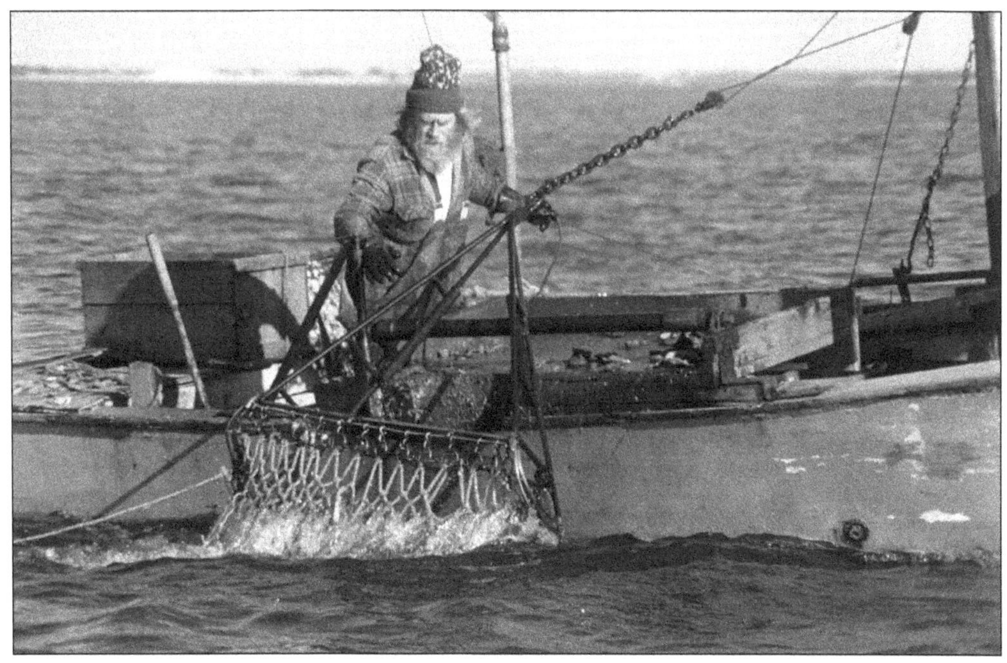

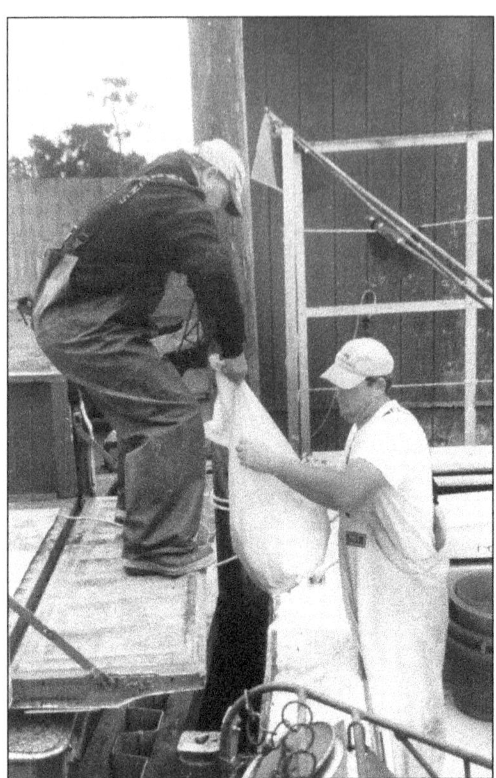

Georgie Daniels, of Wanchese, pictured on the left, went in with his father, Ward, to buy an oyster bed for $3,500 from Cliff Tillett. It proved to be a good investment, and they took countless bushels of oysters from the 50 acres that were delineated by old Buffalo City train tracks. Georgie had to give it up when he was no longer able to put back the amount of shells that the state required. (Courtesy of Georgie Daniels.)

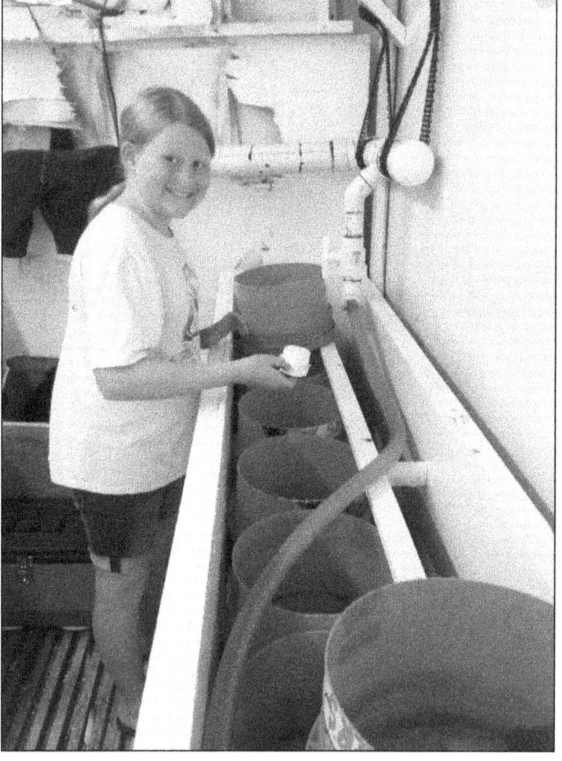

Ten-year-old Emma Daniels introduced oyster larvae into ground shells in a fish-trailer nursery her father built. Joey Daniels presented the idea of farming oysters to the Board of Directors of Wanchese Fish Company, and they approved it. The process began when they leased 50 acres from the state just west of the Bodie Island Lighthouse. Daniels got a crash course in microbiology. (Courtesy of Joey Daniels.)

Pat Leonard broke cages that oysters are grown in out of the icy sound. North Carolina lags far behind other coastal states in leasing acreage and promoting oyster farms. These days, Daniels buys his oyster seed rather than growing his own. It takes from one to two years to get a harvest. The operation will be moving to Hyde County, where the water salinity is more consistent. (Courtesy of Joey Daniels.)

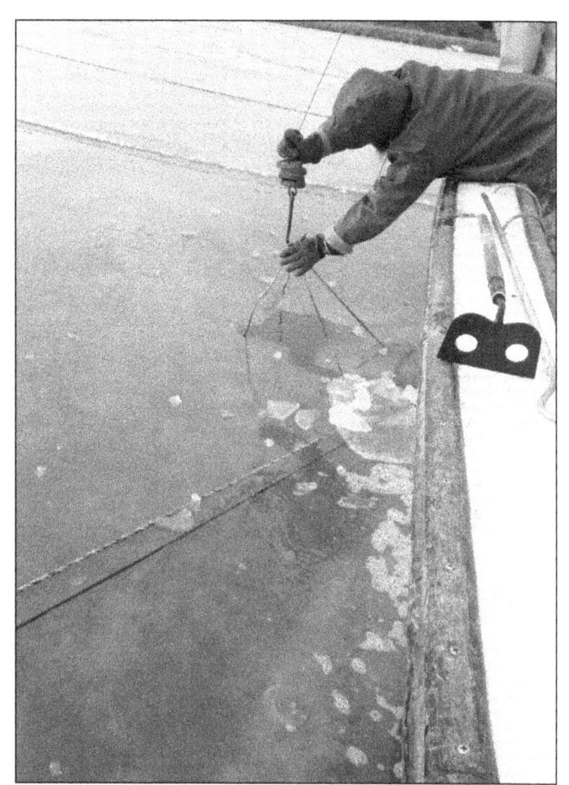

With gnarled hands, fishermen baited a cotton trotline in 1953. This was the way to catch crabs in bulk before the use of crab pots. The men tied a piece of tripe, bull lip, or salted eel every three feet in the long span while curling the line precisely down in a wooden barrel. The barrel had to be loaded just right so the line would flow out after being tied to a gum stake in the water. (Courtesy of Aycock Brown Collection, OBHC.)

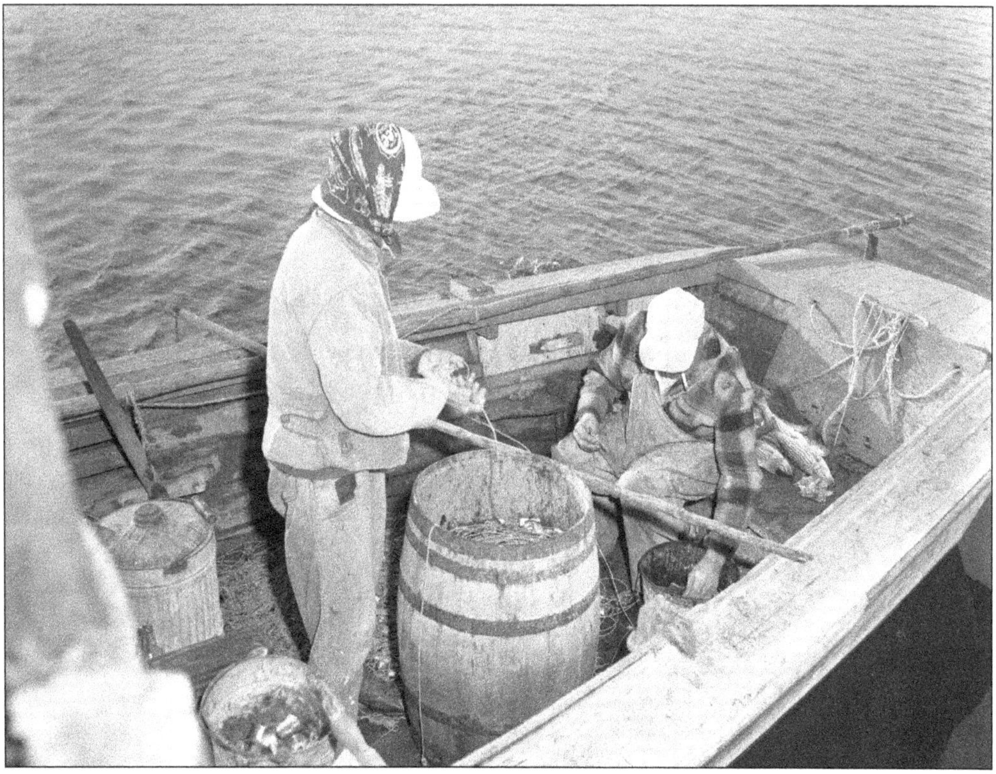

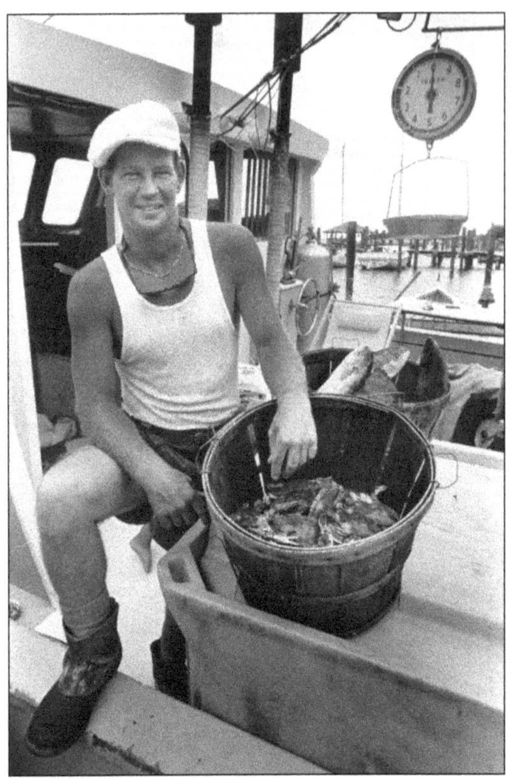

A half basket of crabs was better than no crabs for Tad Dough. The old-time crabbers set their lines before daybreak, hanging a lantern on the starting stake. After setting an ending stake, they would go down the line once or twice before the sun came out, dipping crabs with a dip net, sometimes five or six hanging on each bait, and putting them into barrels. (Courtesy of Drew Wilson Collection, OBHC.)

Since restrictions on crabbing were not as stringent as fishing, many small-scale crabbers operated in the county. The trotline crabber went back to the dock by 12:00 noon because the crabs had sunk to the bottom of the sound to escape the heat. After unloading the crabs, the workers spent the rest of the day at the dock baiting the lines again. (Courtesy of Aycock Brown Collection, OBHC.)

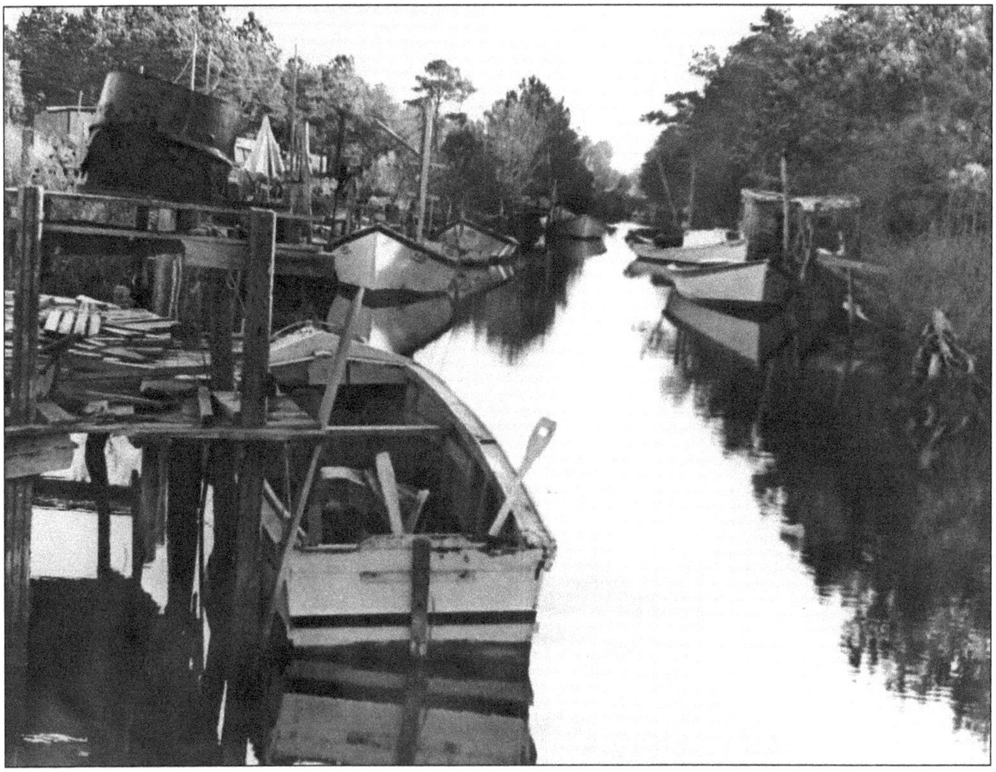

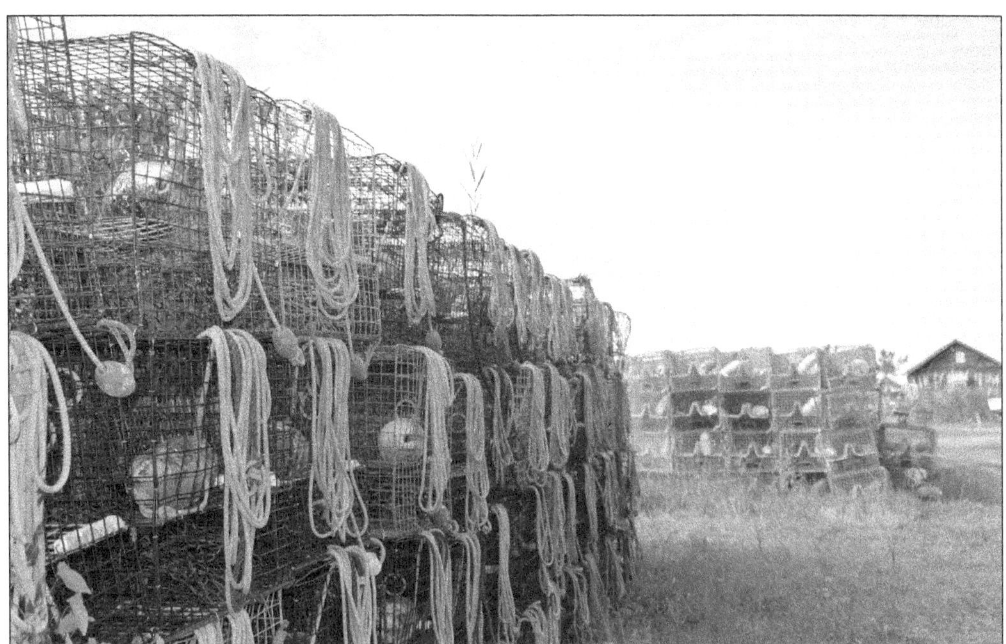

Developed in the Chesapeake Bay area, crab pots were brought into Outer Banks' waters by the 1950s. Each pot has a weight made out of rebar. Pots were initially constructed of galvanized wire, but today's traps are made of colorful vinyl-coated wire, sometimes in the fisherman's favorite team colors. With precisely coiled buoy ropes, these pots are stacked and ready for crab season. (Courtesy of RWG.)

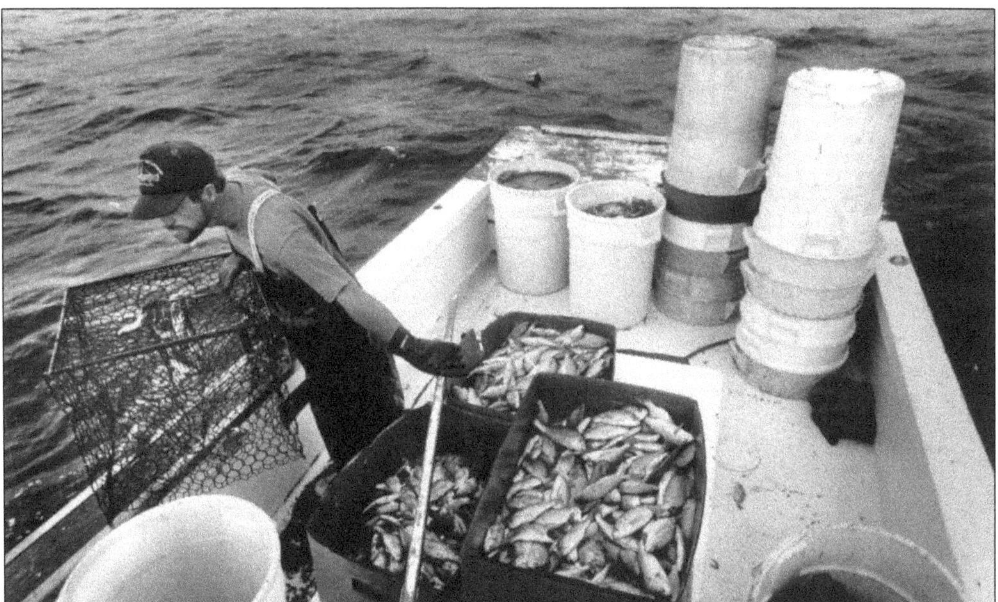

Enticed by the smell of bait, crabs enter the pot through a tunnel and have an option of staying downstairs or going upstairs. Fish are normally stuffed in the bait basket that has a closing door, but many crabbers are opting to leave off the door. Turtles will tear up pots with their beaks, and there is less loss if turtles simply flip the traps to get at the bait. (Courtesy of Drew Wilson Collection, OBHC.)

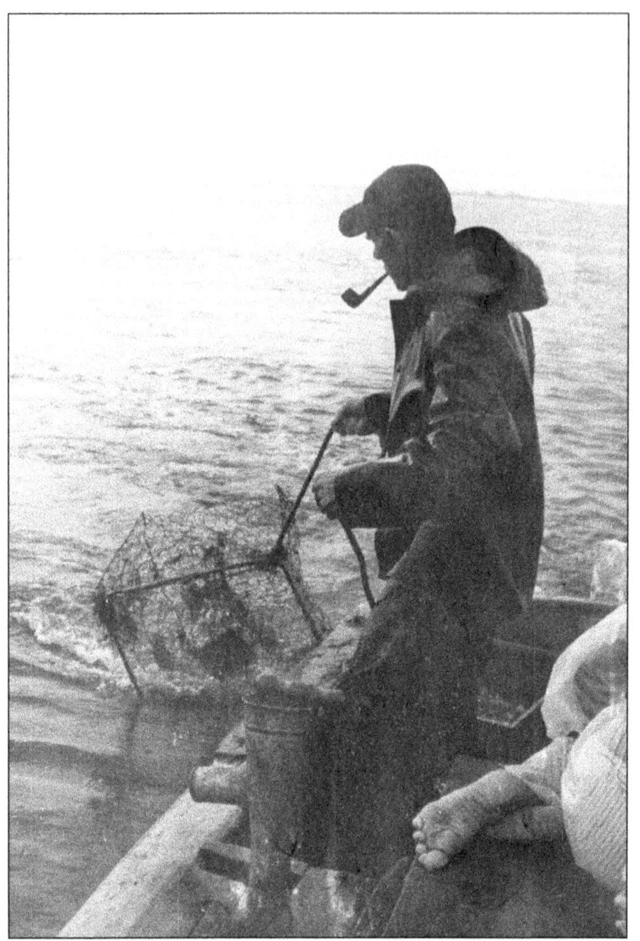

Pulling pots by hand was hard enough, but it took special coordination for Cliff Tillett to do it while smoking a pipe. Tillett's operation was very basic as evidenced by the hand pump in the foreground. Today's crabber has a battery-operated winch to pull pots, enabling him to fish several hundred pots a day by himself. (Courtesy of D. Victor Meekins Collection, OBHC.)

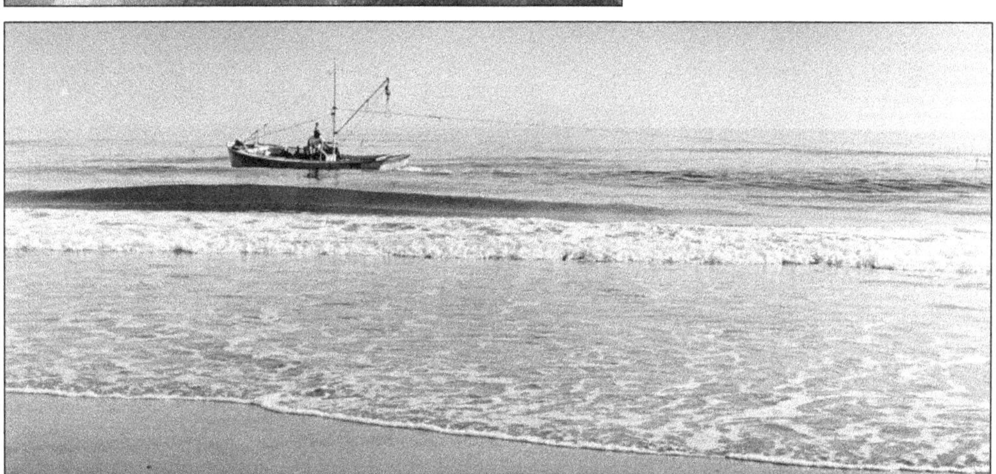

A fisherman pulling a net for crabs just off of a Hatteras beach was an unusual sight. Sometimes dredging for crabs takes place during the winter in the sound when the crabs are hibernating in the mud. Dredged crabs do not fetch a good price because crabs tend to taste like the environment they inhabit. (Courtesy of Dare County Tourist Bureau Collection, photograph by J. Foster Scott, OBHC.)

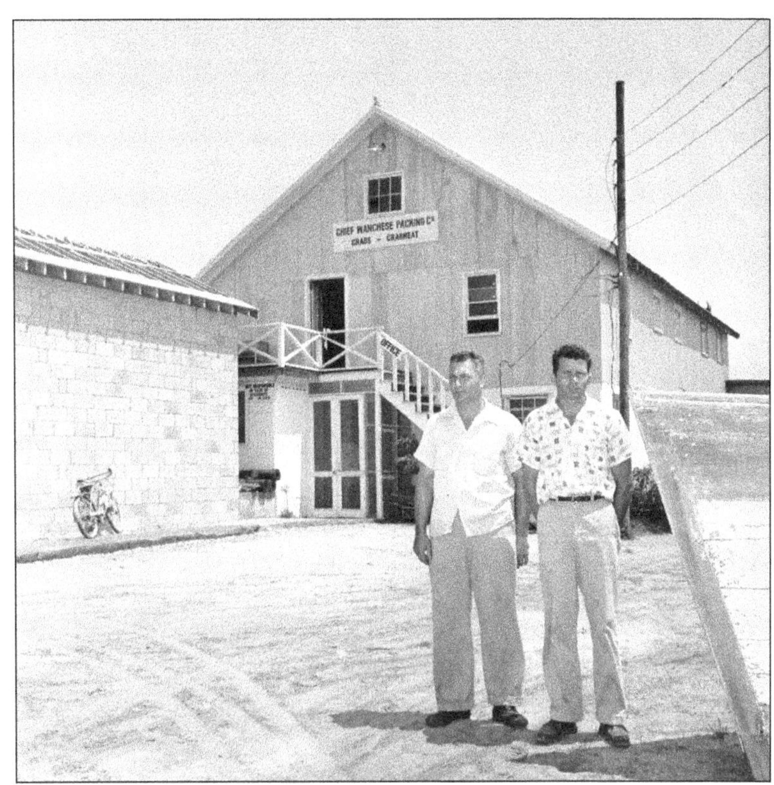

Malcolm Daniels (left) and Ron Tillett (right) partnered in 1957 to establish the Chief Wanchese Crab Packing Company. By the 1950s, shrimp and hard crabs made up half of seafood landings; whereas, in 1890, shad, herring, mullet, bluefish, and oysters accounted for 99 percent of the state's fisheries. These businessmen saw potential earnings in picked crabmeat. (Courtesy of Aycock Brown Collection, OBHC.)

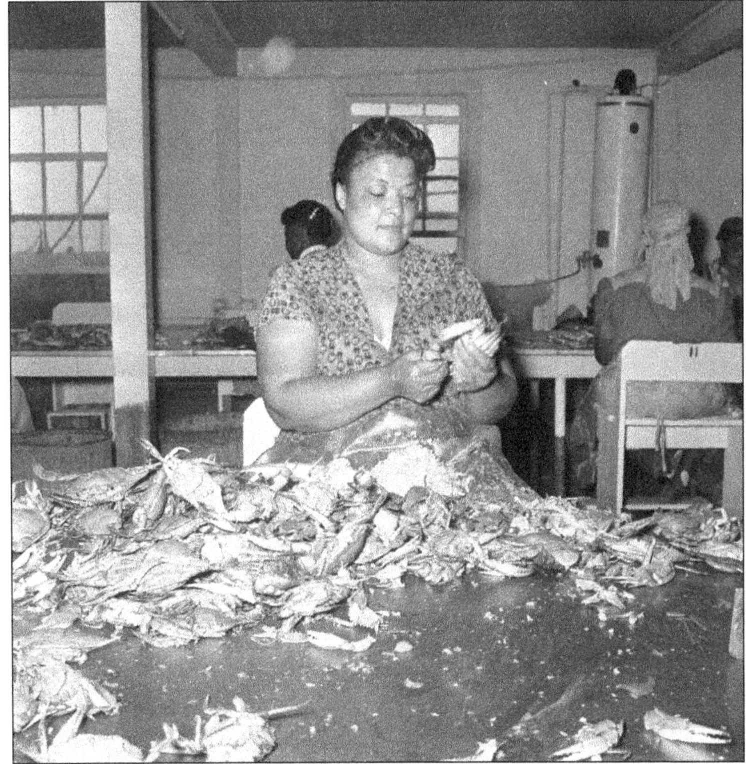

Inside the "crab factory," a picker had an apron full of white crabmeat nestled in her lap and hours of work before her. James Griggs had a crab house near Chief Wanchese. His daughter Lisa remembers as a little girl going from picker to picker, opening her mouth like a little bird, and having each lady place a lump of backfin on her tongue. (Courtesy of Aycock Brown Collection, OBHC.)

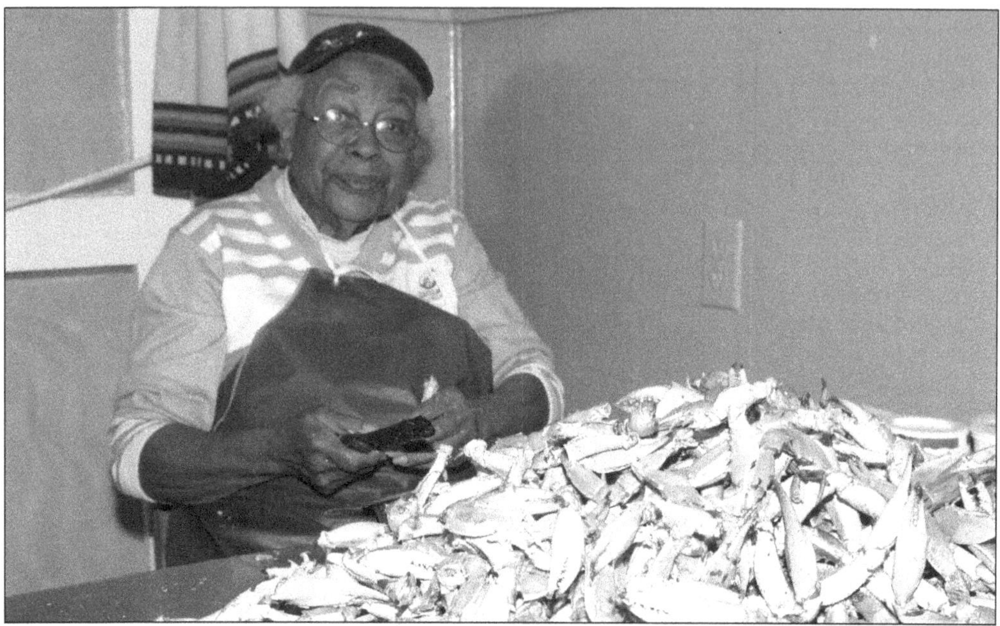

Once a thriving business, Daniels' Seafood at one end of the George Washington Baum Bridge in Nags Head has been torn down. Owner Mickey Daniels would pick up his employees in a van every day but Sunday. The ladies sat around a large table piled high with steamed crabs, picking, talking, laughing, and singing hymns together as if they were at a church social. (Courtesy of Lisa Griggs.)

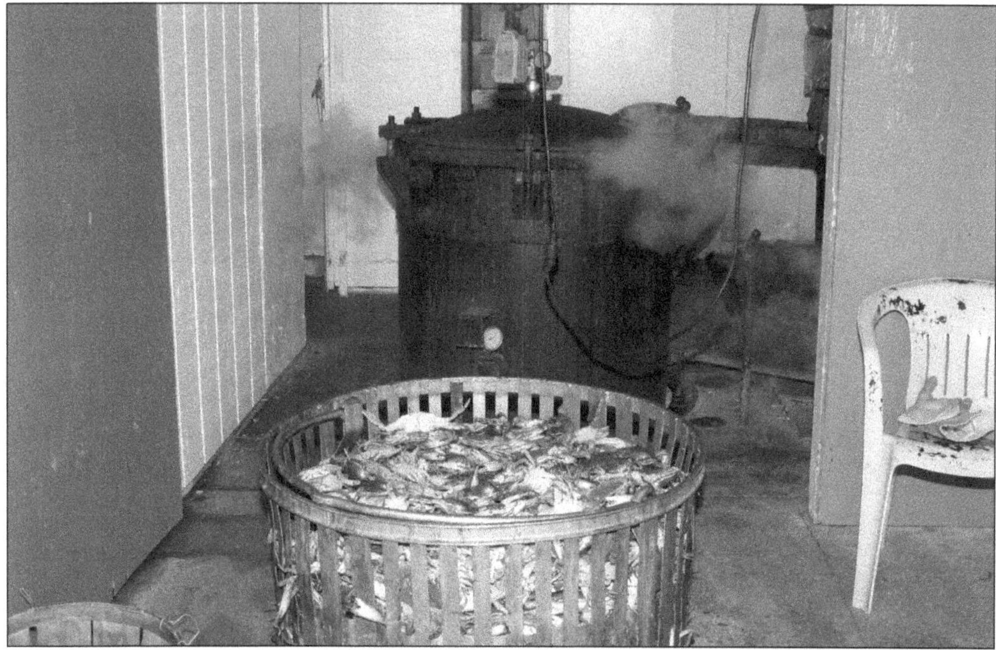

The scientific name for a blue crab is *Callinectes sapidus* Rathbun. The Greek word, *Callinectes*, means beautiful swimmer, and the Latin word, *sapidus*, means tasty. Rathbun is the Smithsonian scholar who specifically identified 998 crabs, giving only the blue crab a name that included its gastronomical quality. Blue crabs will turn red when steamed. (Courtesy of Lisa Griggs.)

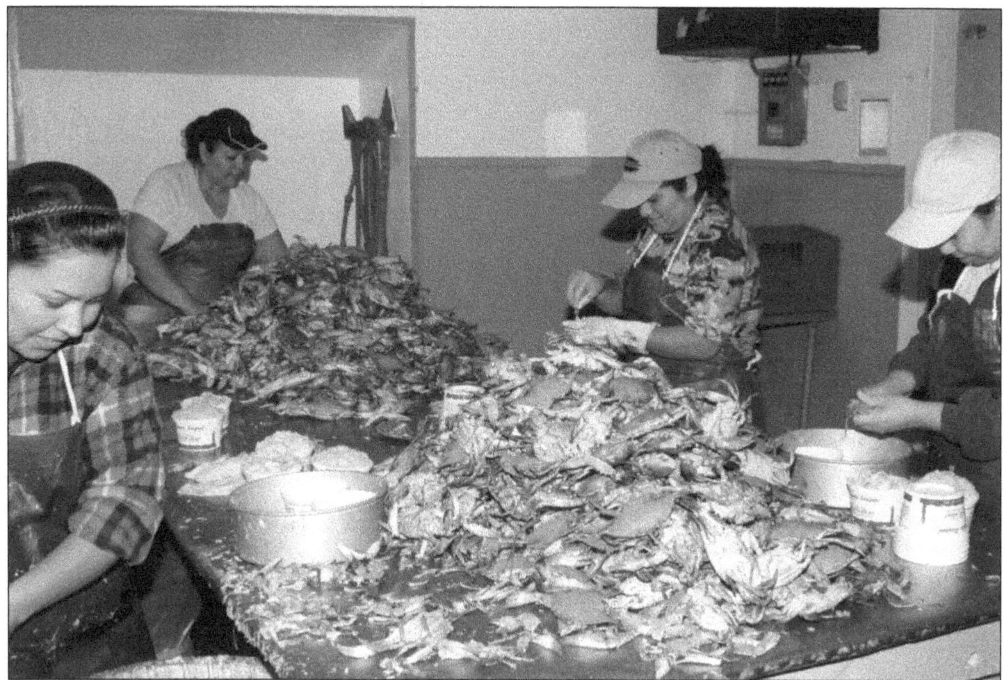

Handpicked crabmeat is far superior to a mechanically obtained product. It takes 100 pounds of crabs to get seven to eight pounds of pure crabmeat. There are no longer any local crab houses, due in part to strict health department rules and a labor shortage. Before closing, women from Mexico came as migrant workers to work at Daniels' Seafood for the season. (Courtesy of Lisa Griggs.)

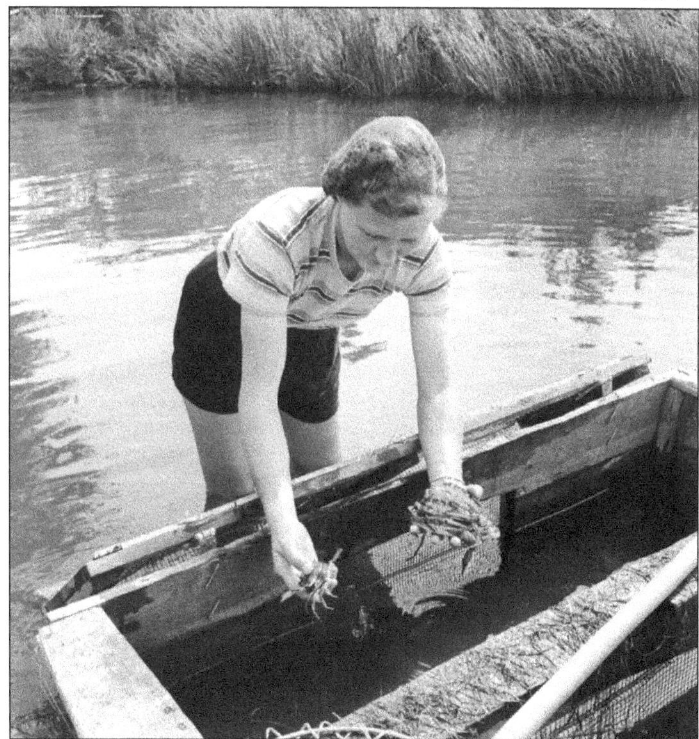

In 1956, a Colington woman retrieved handfuls of "busting" crabs from a homemade shedder in a canal. Each spring blue crabs molt in preparation of mating and are temporarily rendered soft while new exoskeletons harden. Enterprising locals waded along the sound side to find crabs ready to shed, and they retained them until the crabs were soft. (Courtesy of Aycock Brown Collection, OBHC.)

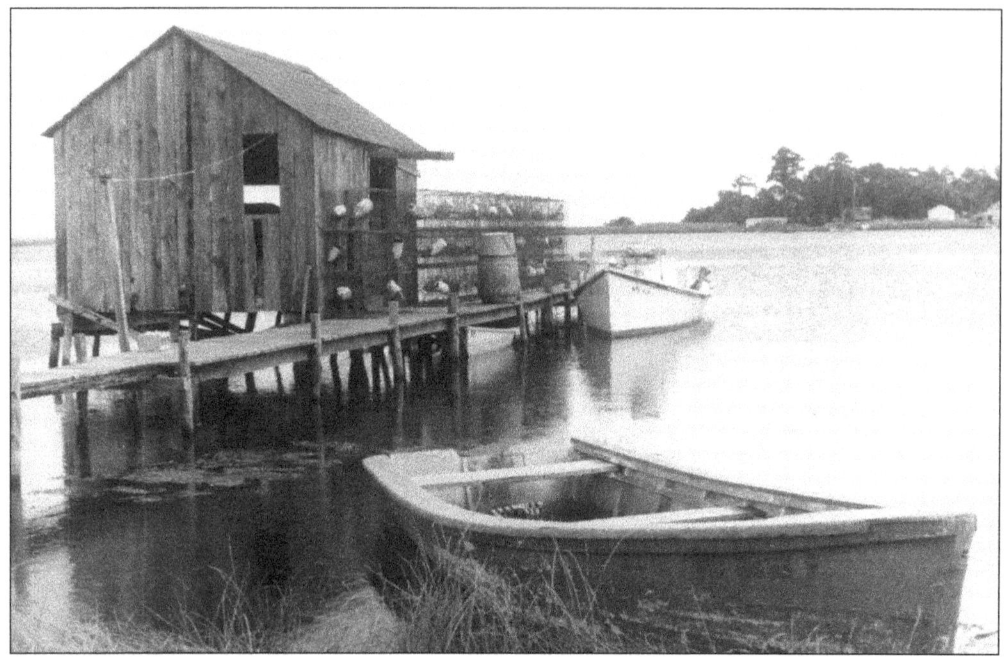

Old methods are eventually replaced, just as this unused boat was pulled out of the way at a Colington dock. Murray Bridges saw the future of commercial soft crabbing in the early 1970s. Taking his Endurance Seafood to production level, he has 150 land-based, shallow-tank shedders plumbed with circulating water to maximize yield during a brief window of time. (Courtesy of Aycock Brown Collection, OBHC.)

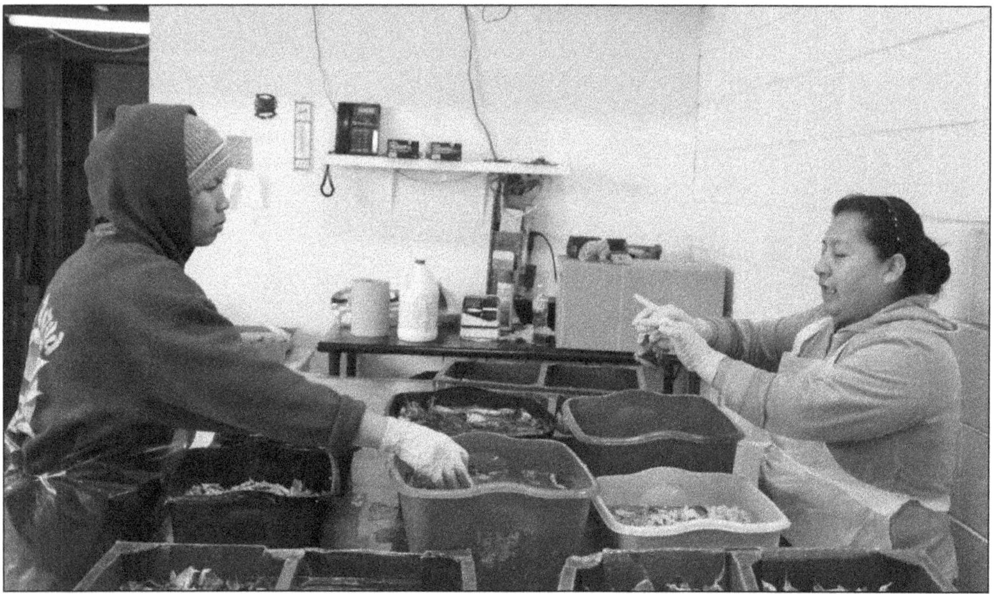

Two stories of indoor shedders occupy O'Neal's Sea Harvest. The owners continue shedding crabs beyond the blockbusting April-to-June mating season by accepting "peeler" crabs that are caught all summer long in lesser quantities. The majority of soft crabs are sold live, but cleaned, wrapped, and frozen crabs are marketable and prepared by these two ladies originally from Thailand and Mexico. (Courtesy of RWG.)

Seven

Gearing up for Change

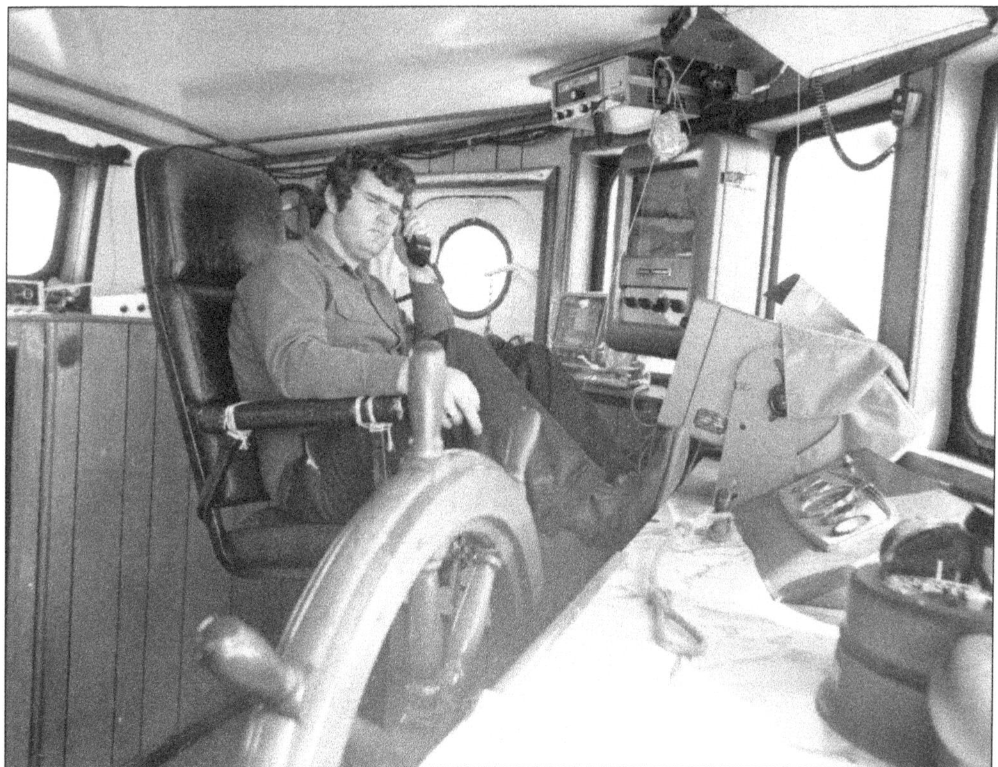

If any fisherman knows about being adaptable, it is Billy Carl Tillett. He tried his hand at squidding in 1983 when urged by a New Jersey fish dealer. He gave it a shot, adding a net liner to his existing shrimp net. On the second day, in one drag, he pulled up 60,000 pounds of glistening squid. He became a believer in exploring new fisheries. (Courtesy of Billy Carl Tillett, photograph by J. Foster Scott.)

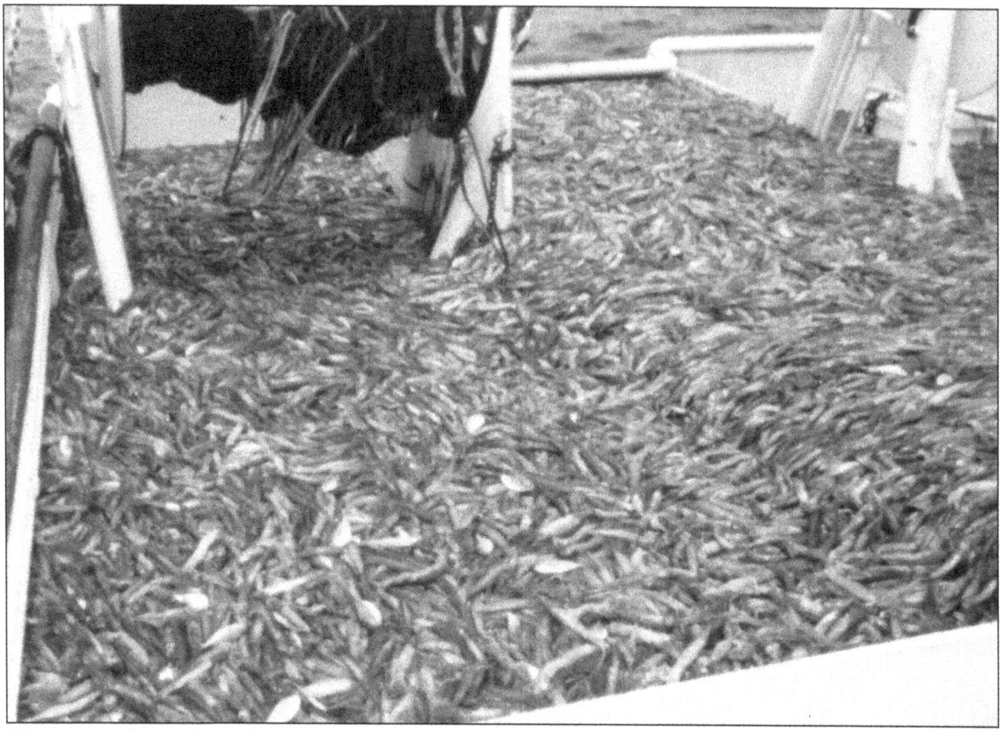

Winter squid, or *loligo*, almost spilling over the sides of the *Linda Gayle*, represented a huge payoff at the going price of 15¢ per pound in the 1980s. Tillett explained that some years he could not catch them since they migrated in 100 to 1,000 fathoms of water and "had a lot of space to run." Catches like this one made Tillett wish for an onboard blasting freezer. (Courtesy of Billy Carl Tillett.)

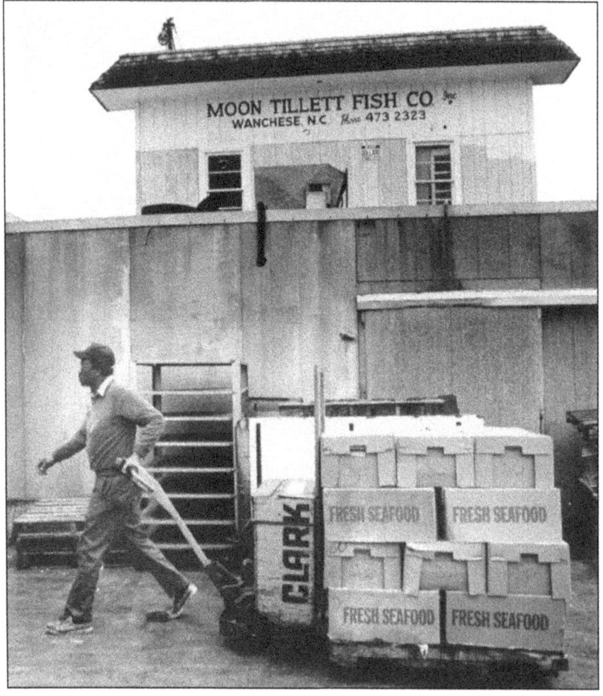

Dock worker Sam Weston hustled along with a motorized pallet jack and a load of squid in March 1989. The Tilletts kept their freezers packed with the squid they caught with their own boats. They exported squid to China, experimenting with loading their own shipping containers until the logistical arrangements became too complicated. (Courtesy of Drew Wilson Collection, OBHC.)

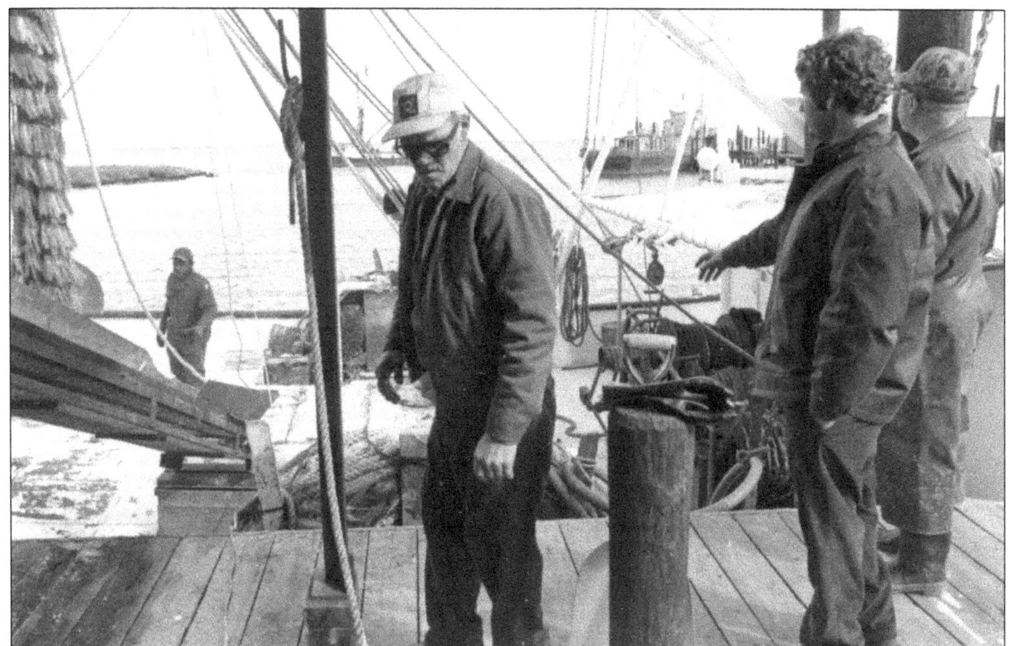
In 1983, between July and October, the Tilletts caught over a million pounds of squid. Coauthor Wayne Gray remembers when squid was shoveled off the docks, unused and unwanted. Willie Etheridge III talked Gray into serving it as a calamari appetizer at his restaurant, Queen Anne's Revenge. It became a customer favorite. (Courtesy of Billy Carl Tillett, photograph by J. Foster Scott.)

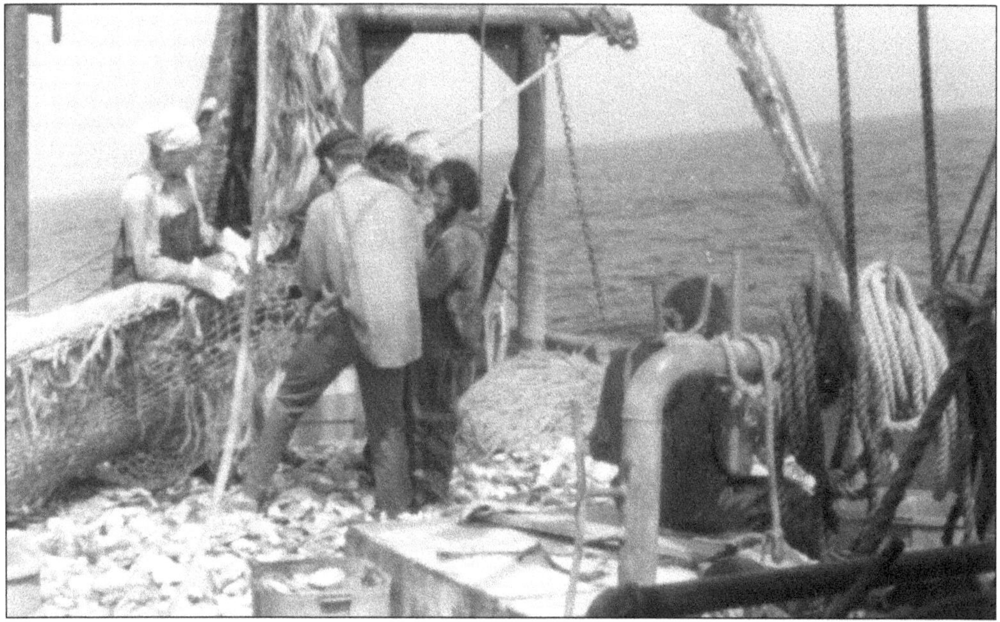
When Frog Tillett (center) first brought in scallops as a bycatch in 1962, he had no idea what to do with them. When boats from New Bedford, Massachusetts, began dredging beds off of North Carolina, locals started gearing up. A little more than a decade later, Tillett would be the successful captain of the scallop boat, *Vickie,* cashing in on one of our nation's most valuable industries. (Courtesy of Frog Tillett.)

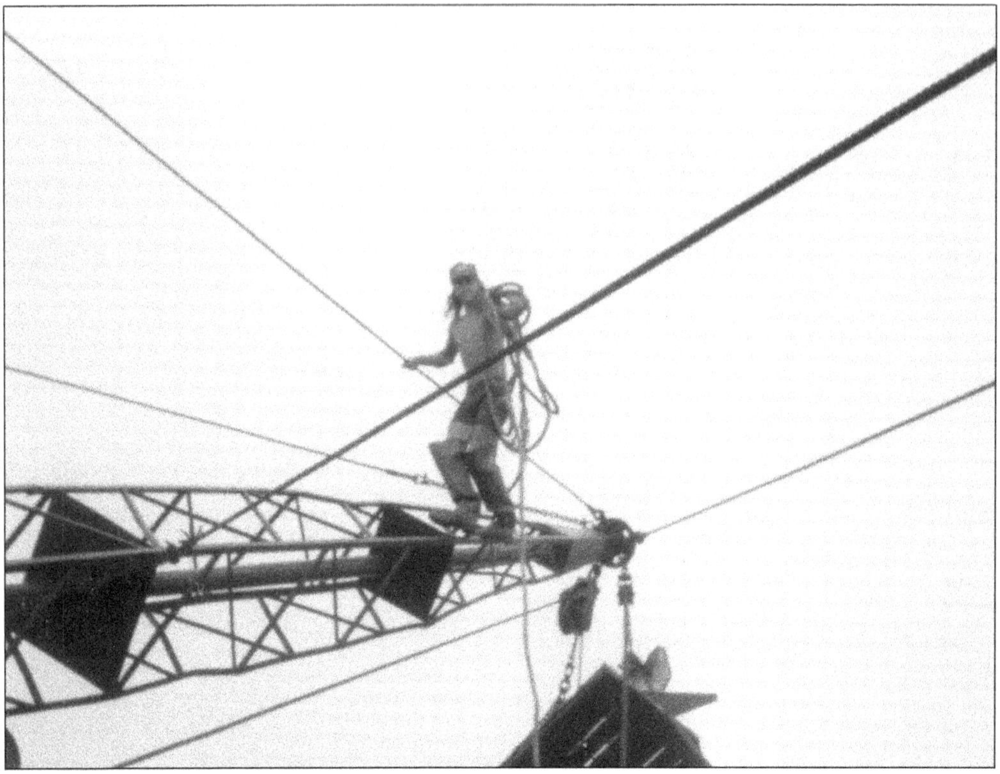

Crewmen have to be versatile and willing to do whatever it takes to get the boat and equipment operational. Working on motors and hydraulics, untangling ropes from propellers, and loosening bound chains are just a few emergency jobs that might be needed on a trip. In this instance, "Little Jeffrey" Midgett climbed the rigging to untwist cables for the doors of a scallop rig. (Courtesy of Jeffrey W. Midgett.)

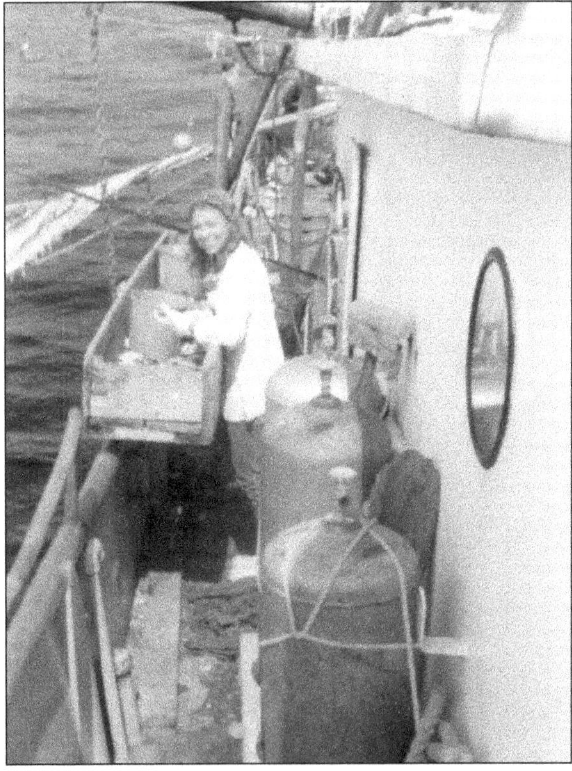

Linda Pisciotti (pictured) and her girlfriend Michelle Cindrich came to the beach from Pennsylvania in 1976 looking for summer adventure. They got that and more when they signed on as cooks aboard the *Vickie*. Each crewmember had to help shuck the scallops, cutting out the edible adductor muscle, the strong muscle that opens and closes the shell. (Courtesy of Frog Tillett.)

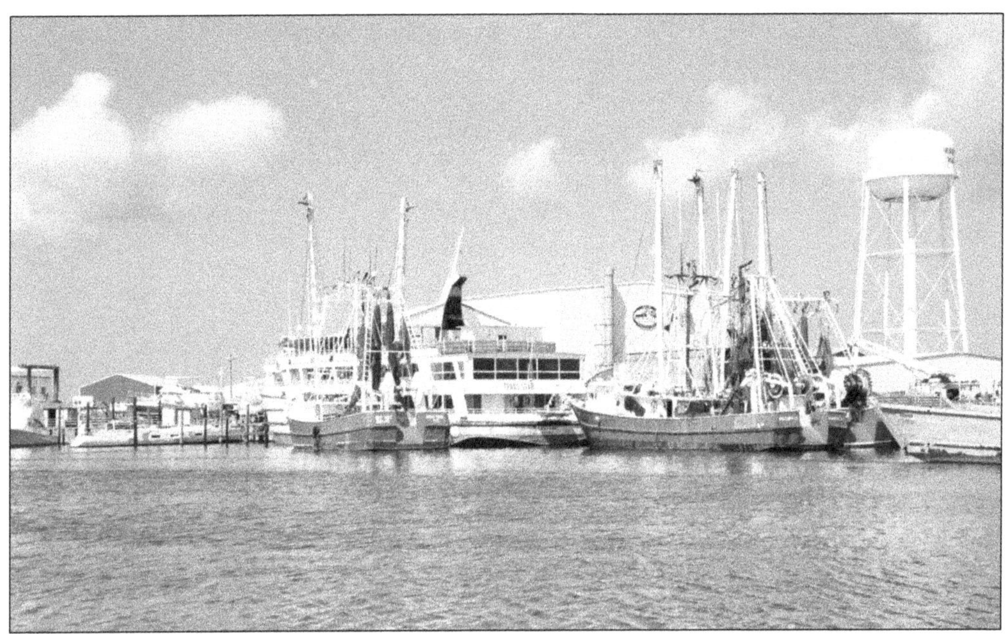

Wanchese Fish Company acquired the former casino, *Texas Star*, when calico scallops were plentiful and federally unregulated. It took four years and untold dollars to reconfigure it to catch, shuck, and freeze the product. Finally, *Texas Star* was completed and headed out to sea in the fall of 2014. To the great disappointment of those involved, the small scallops could no longer be found. (Courtesy of Lisa Griggs.)

Comparatively speaking, Duck residents were late in getting electricity, phone service, and paved roads. The area was too primitive for most tourists. Besides a naval outpost, fishing was the only industry. In 1955, these women and men worked together to mend net. Tourism rules the economy of Duck now. There are no fishing families remaining. (Courtesy of Aycock Brown Collection, OBHC.)

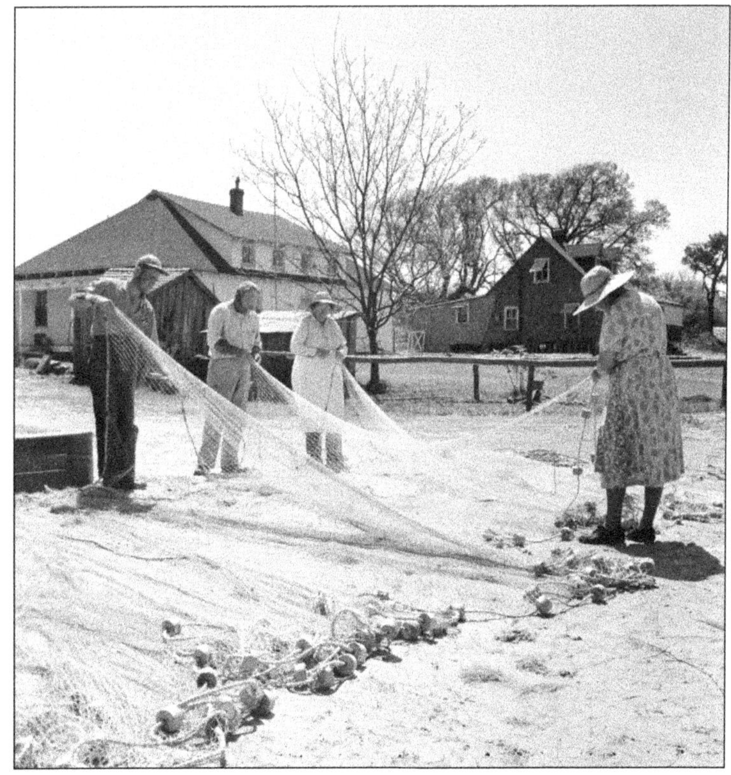

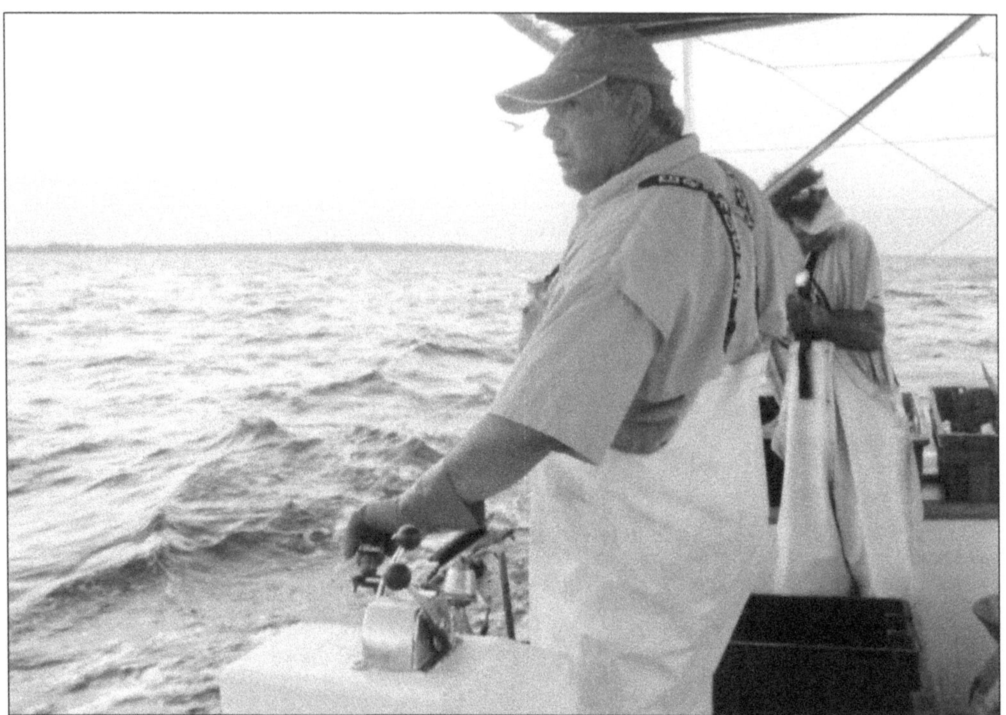

A life of independence has a cost, especially when a man is no longer able to work. When Tracy Payne developed a heart condition at a young age, there was no framework of support that might have been accessible in other types of work. More often than not, a fisherman's retirement plan comes down to the sum of money he gets when selling off his boat and gear. (Courtesy of Tracy Payne.)

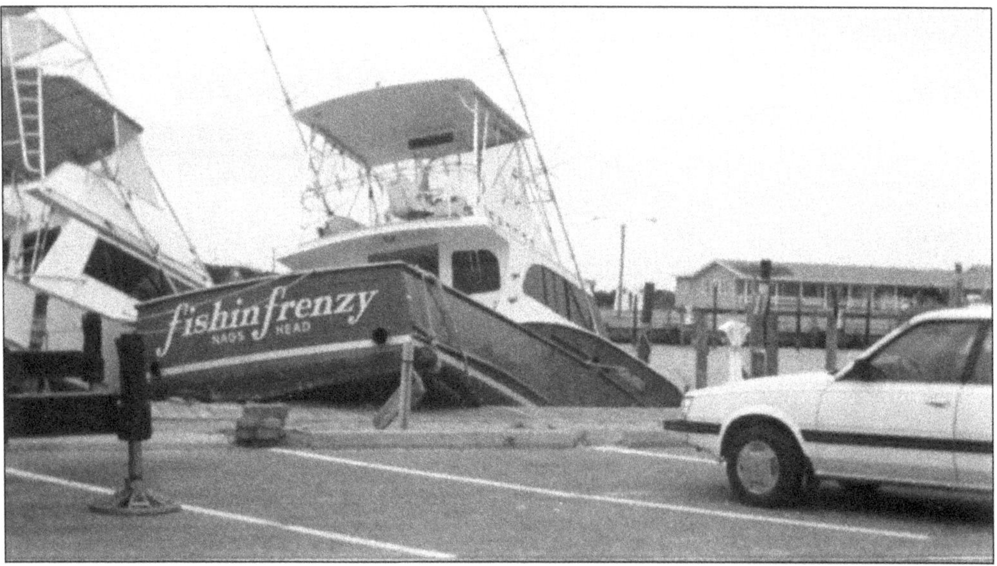

Battling the elements of nature takes place on the water and, sometimes, at the dock. A hurricane tossed part-time commercial vessel *Fishin Frenzy* up on the Oregon Inlet Fishing Center dock like a toy boat. Maritime insurance is pricey, and quotes are based on a variety of factors, such as the age of the vessel, type of engines, mooring location, and intended area of navigation. (Courtesy of Tracy Payne.)

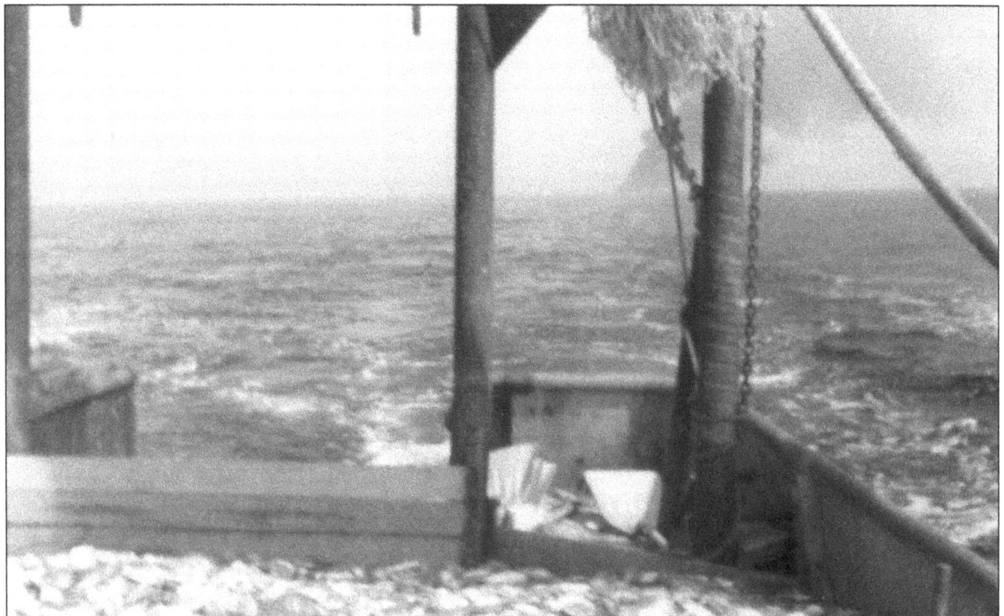

When a new Methodist minister came to Wanchese, he proposed erecting a memorial to men lost at sea like those seen in New England fishing villages. The community concluded that it has been blessed, and a memorial was not needed. Despite numerous perils, there has been very little overall loss of life. In the late 1970s, the burning *Holly F. Murphy* was photographed from the deck of *Vickie*. (Courtesy of Frog Tillett.)

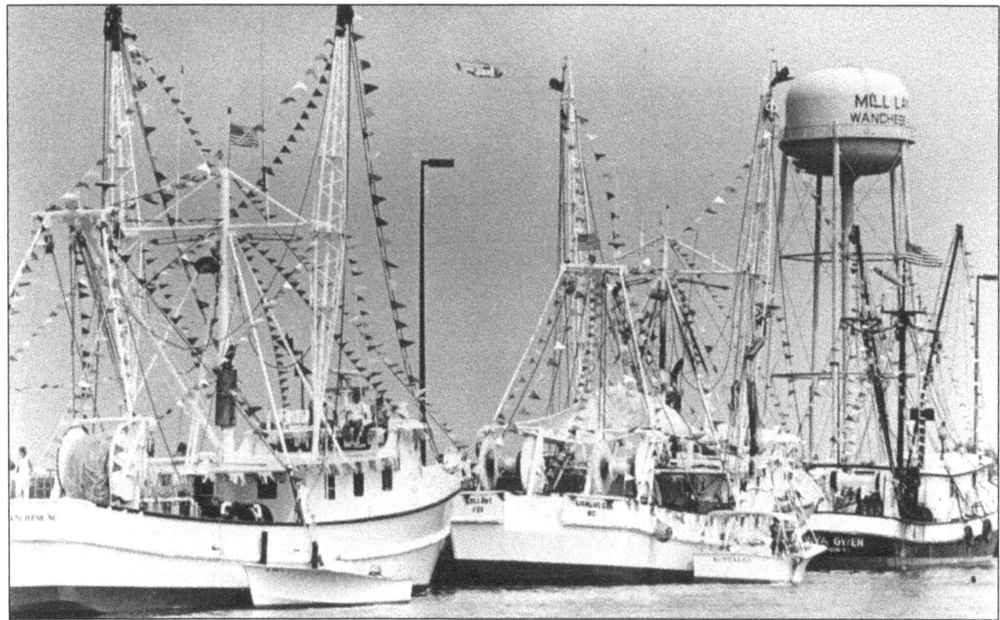

Spruced up to celebrate, trawlers line up for the Wanchese Seafood Festival in 1987. These days, Hatteras Village honors working watermen by hosting "Day at the Docks" each fall. It is a day of fun with contests, races, and cooking demonstrations. It is a day of education with field programs and boat tours. It is also a day of reverence with the blessing of the fleet. (Courtesy of Drew Wilson Collection, OBHC.)

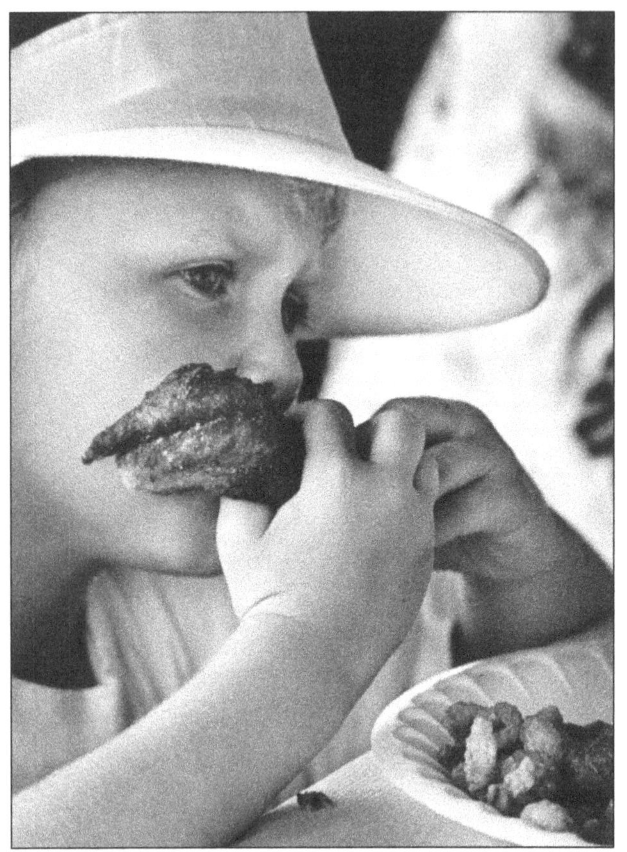

Four-year-old Kate Herron of Waynesville, Virginia, chomped on a Spanish mackerel fillet at the 1991 Wanchese Seafood Festival, which was attended by about 2,500 visitors. Currently, the Outer Banks Seafood Festival has been held in Nags Head since 2011. Its mission is to promote local seafood and restaurants while proclaiming the area's coastal heritage. (Courtesy of Drew Wilson Collection, OBHC.)

Thanksgiving time used to be the biggest week of flounder fishing, and when wooden trawlers first ventured far out in the ocean, it was fluke they were after. Flounder are classified as a depleted resource, but the lucky few who hold permits argue that their quota of 5,000 pounds can be caught in less than 48 hours, making them wonder just how many flounder really are on the ocean floor. (Courtesy of OBHC.)

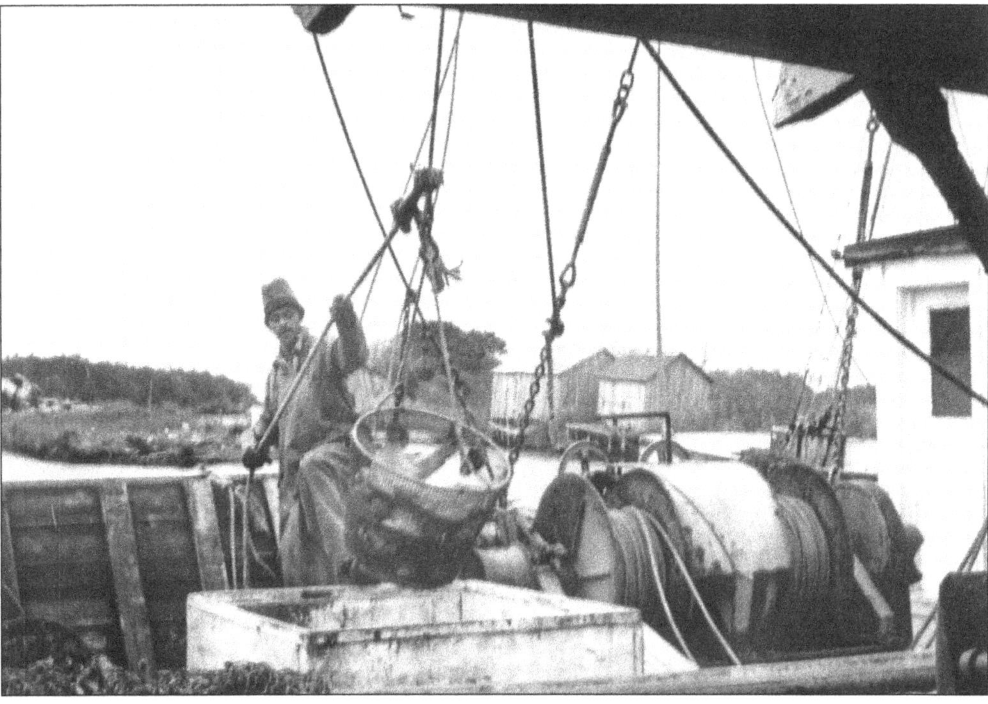

In 1953, a man, identified only as "Gray," held up a flounder that could feed a whole family. Whether it is natural cycles or whether it is climate change from global warming, some fishermen have observed that all fisheries have moved about 200 miles north. A boat must steam to New Jersey to get the flounder that were once so plentiful off of North Carolina. (Courtesy of Aycock Brown Collection, OBHC.)

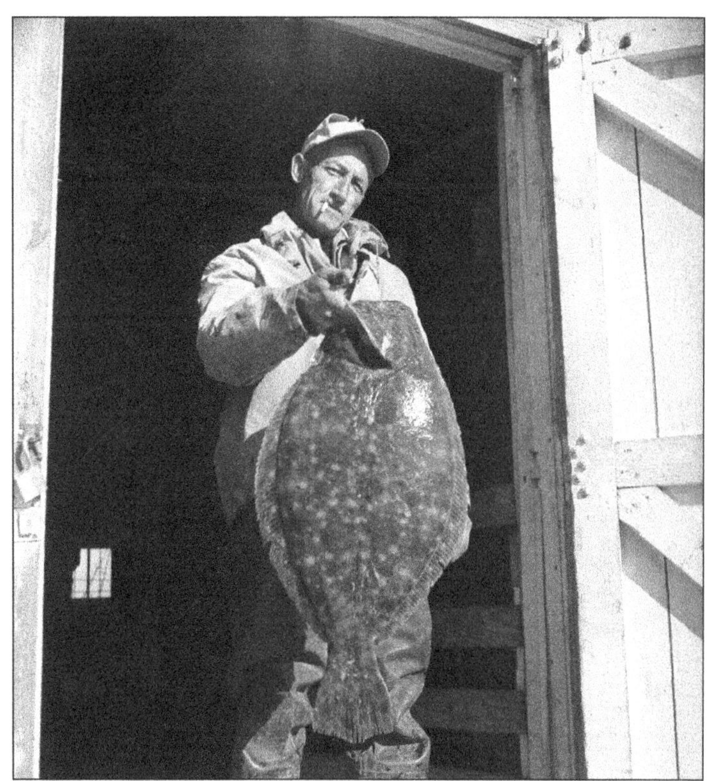

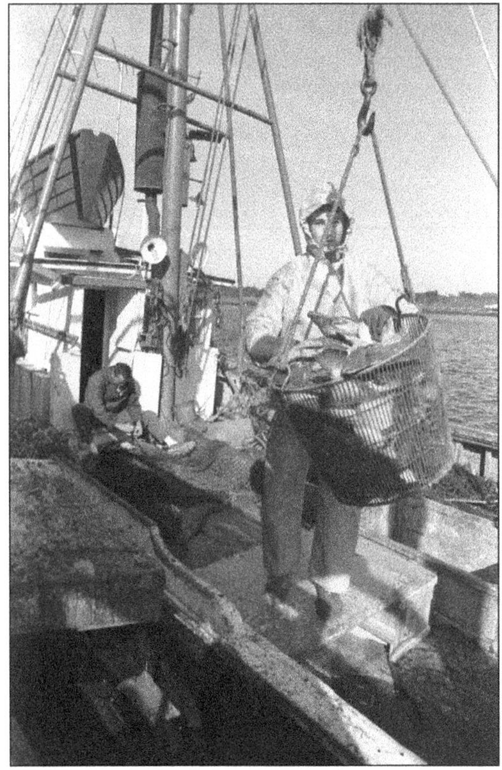

In pre-regulation days, a young man handles a swinging basket of flounder brought up from the hold. Seafood dealer Mark Vrablic holds the view that landing statistics are skewed. He contends that flounder fishermen go elsewhere to catch their quota because they do not want to deal with the expensive turtle exclusion devices that North Carolina requires. (Courtesy of Bruce Roberts Collection, OBHC.)

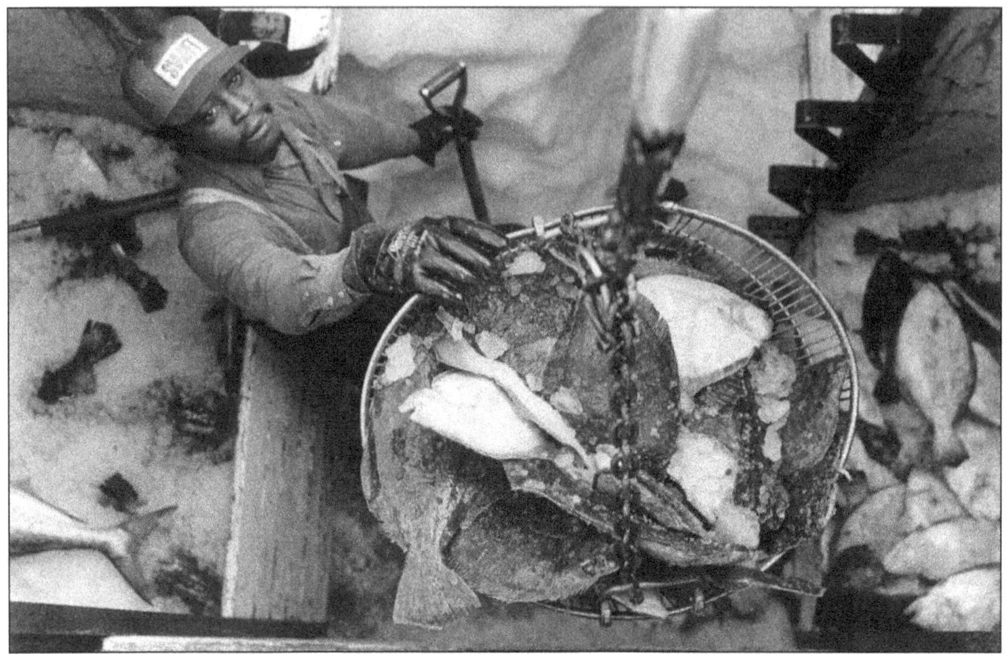

Steadying the basket, this man did not want to spill any of the flounder he just released from an icy tomb. Most watermen agree with elder Omie Tillett, who says that mankind needs laws and regulations. However, there is a level of frustration when scientific data does not match up with what they see and experience daily on the water. (Courtesy of Drew Wilson Collection, OBHC.)

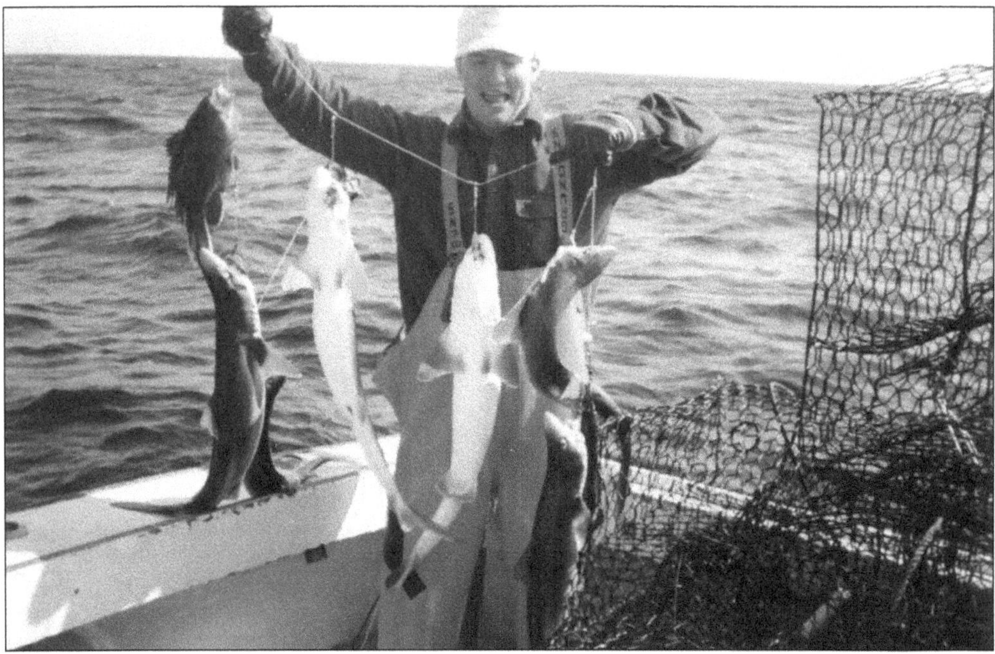

Britt Shackelford hollered with delight over a string of dogfish. When the industry's focal point turned from local to foreign markets, there was no longer any such thing as trash fish. The boneless spiny dogfish were shipped to Massachusetts where they were filleted, boxed, and flown out of the Boston airport to either England, Germany, or Belgium. (Courtesy of Britton Shackelford.)

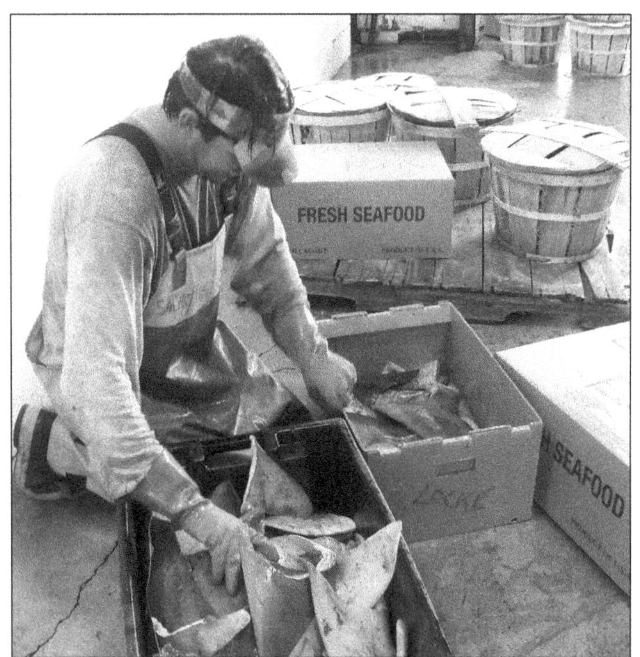

Charlie Locke carefully handled his shark fins that were most likely headed to Asian markets. Although the taste of fins is neutral, shark fin soup is an emblem of status because it is time consuming and complex to make. Tails and swimming fins are used in a less-expensive version of the Chinese soup. (Courtesy of RWG.)

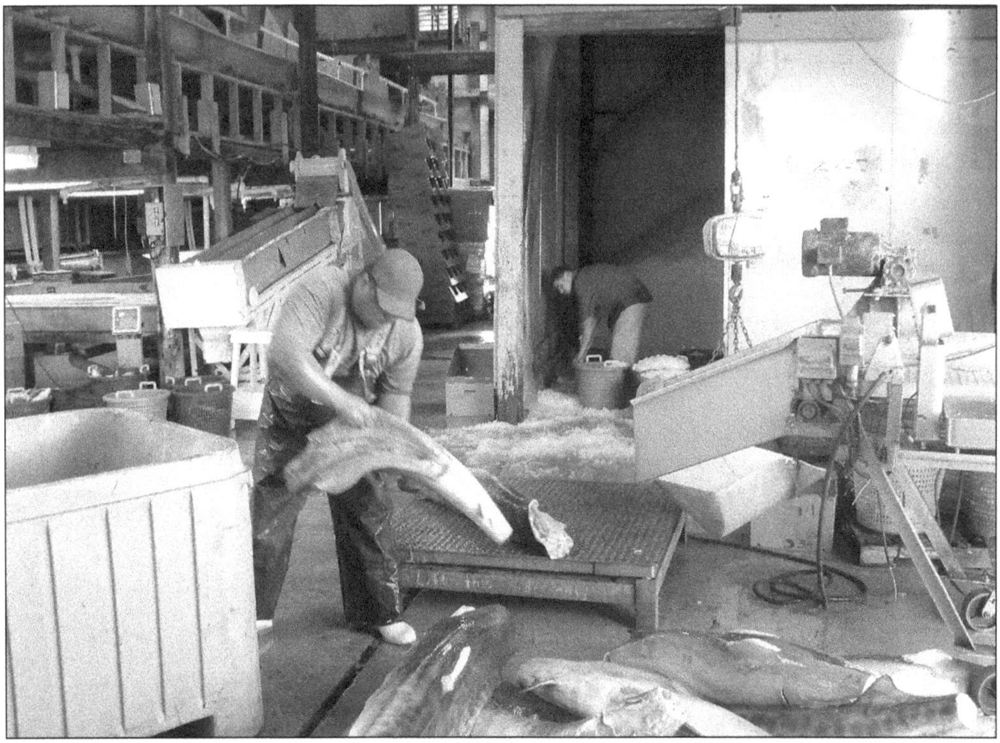

In November 2017, many different kinds of sharks were laid out for Jose Gonzales to weigh on floor scales and pack in insulated vats at O'Neal's Sea Harvest. Fishermen often shoot sharks with rifles before bringing them on board. Sharks thought to be dead have bitten more than one man. A best seller in Asia, Mexico, and Africa, shark meat is slow to gain popularity at home. (Courtesy of RWG.)

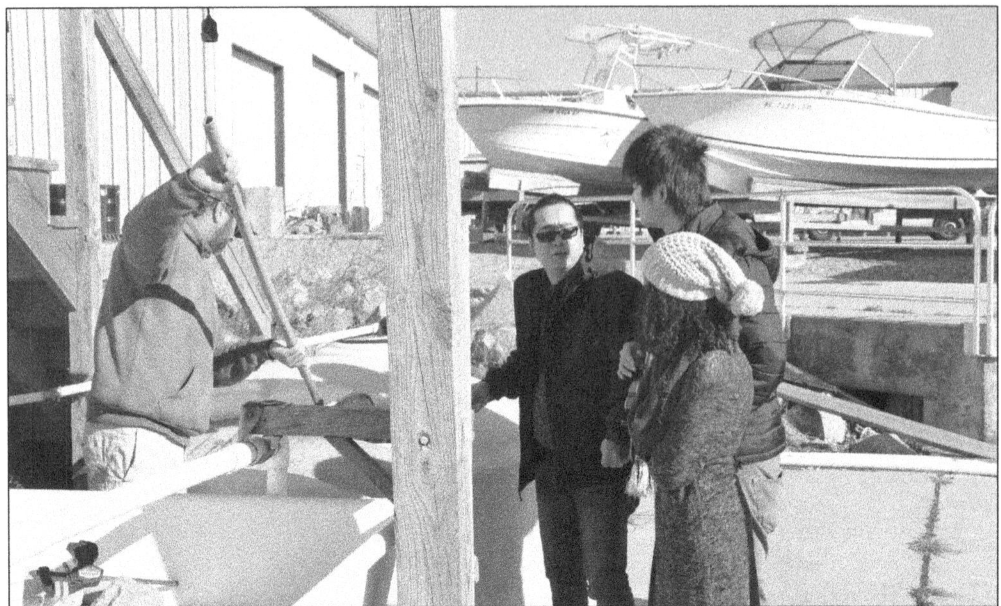

Fisherman Colby O'Neal (left) had to take a hard stand with connoisseurs who wanted him to try and dip a specific wriggling eel out of bin full of its writhing brothers. "You get whatever I dip up," he growled. O'Neal has capitalized on a seasonal niche market, catching and shipping eels north, where many families eat them traditionally for Christmas and New Year's Day celebrations. (Courtesy of RWG.)

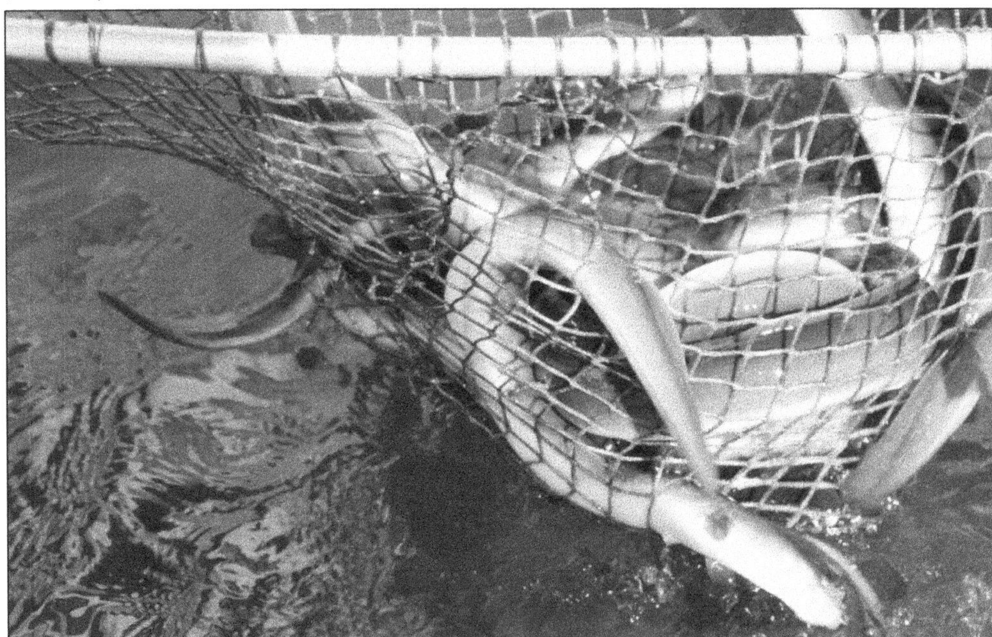

Eels are amazing creatures that live most of their lives hiding in seaweed in an area in the Atlantic that is bordered by four currents. They are caught in pots when they come to coastal estuaries to reproduce. They are smart enough that, when put in a barrel, they swim around the circumference quickly, building momentum to surge up, hang their heads over the rim, and fling their bodies over to freedom. (Courtesy of RWG.)

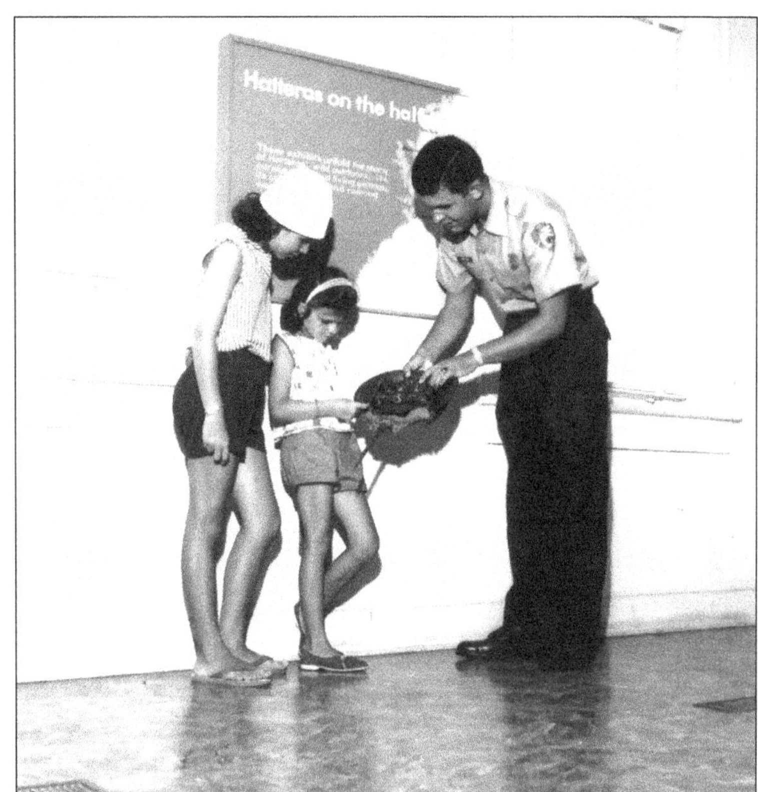

In 1963, National Park Service naturalist Jerry Stone explained the wonders of the horseshoe crab at Bodie Island Visitors Center. Live horseshoe crabs are shipped to laboratories where they are bled and released. Limulus Amebocyte Lysate, derived from their blood, is used to test intravenous drugs and medical equipment for the presence of bacteria. No manmade substitute is as effective. (Courtesy of NPS, CHNS.)

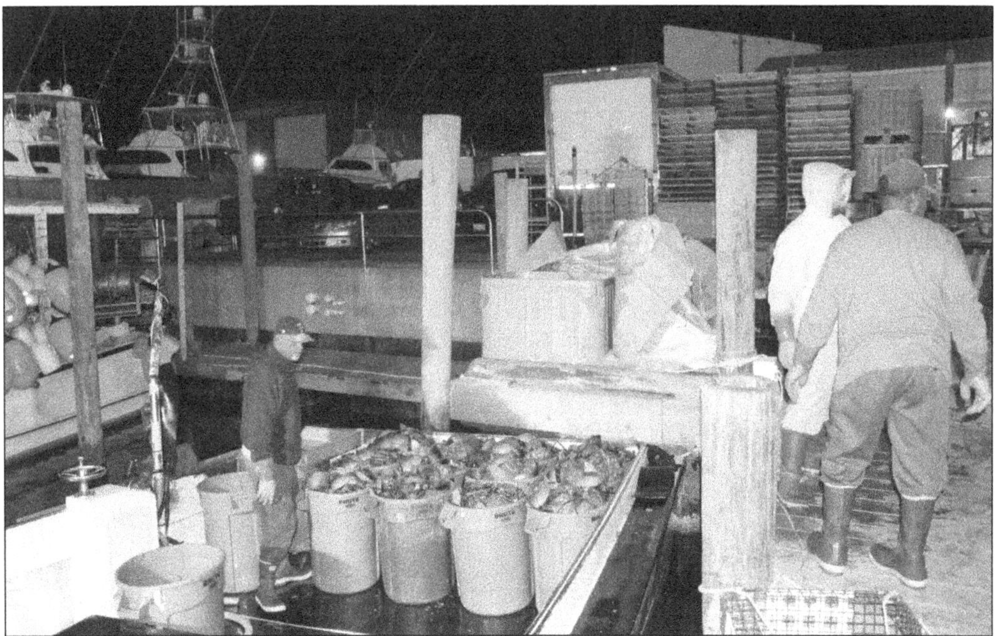

Barrels of horseshoe crabs that will be used for bait are loaded on a conch boat making a quick turnaround. The arthropods are highly regulated, in part, because a shorebird, the red knot, feeds exclusively on their eggs. Red knots take a layover in the Delaware Bay during their migratory flight from South America during horseshoe-crab mating season. (Courtesy of RWG.)

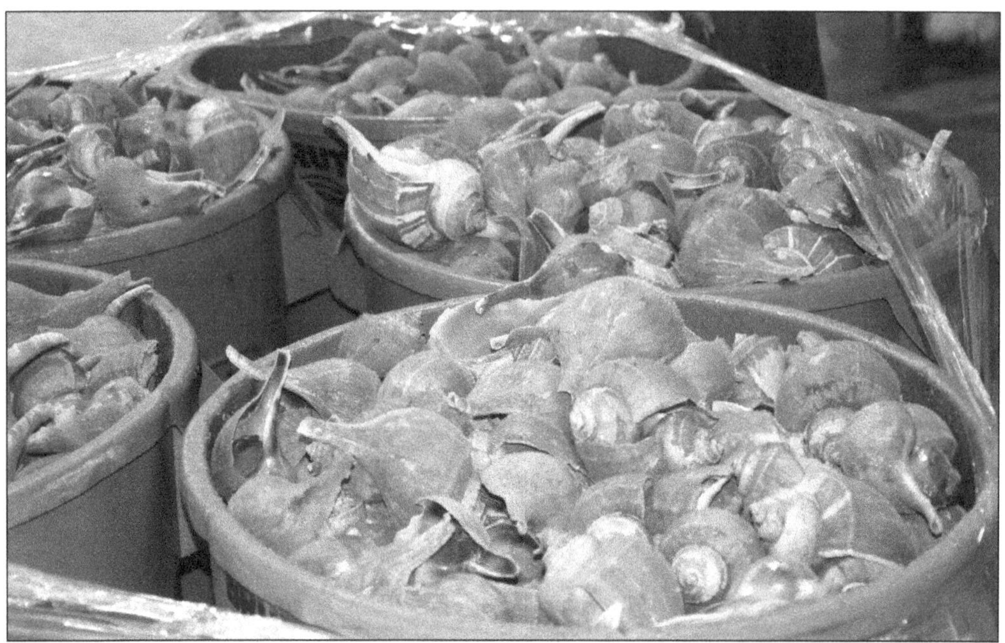

A serious beachcomber would love to have just a few of the thousands of ocean whelks that a winter conch fisherman trapped off of Corolla in 2018. Each conch trap is baited with a horseshoe crab impaled on a spike. The channeled whelks pile in to feast but come to a standstill when leaving as they encounter a rim around the otherwise open top of the trap. (Courtesy of RWG.)

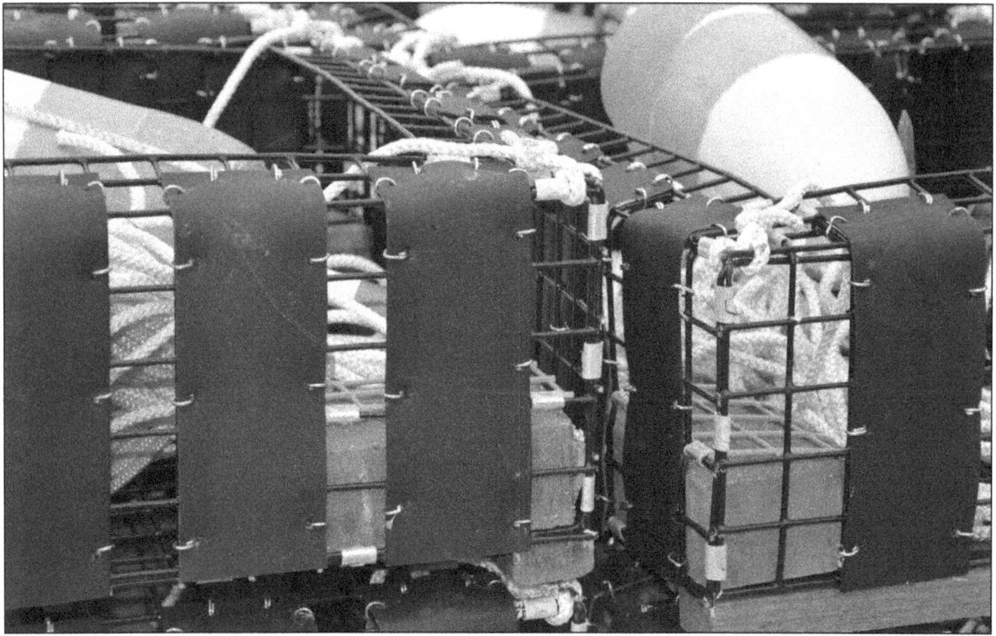

Catching whelks is another example of ambitious fishermen trying to attain an unregulated product. A homemade pot has a brick for weight and ropes configured like a harness to keep the pot level when lifted. Trawl lines are made up of 20 pots on 1,000 feet of rope, and are set behind the breaking surf. It is a lucrative small fishery with the fish house paying about $2.90 per pound with the shell. (Courtesy of RWG.)

Eight

Big Fish, Big News

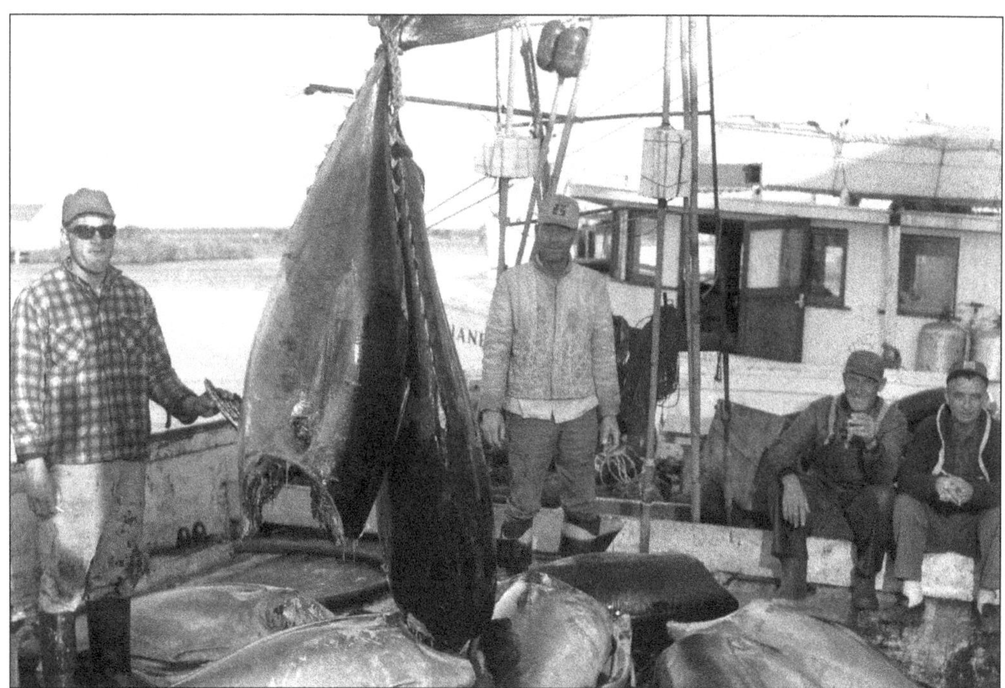

When swordfish longlining was introduced locally in 1961, a race to riches began as Wanchese boats rigged up for a new opportunity. James Griggs (far right) and his vessel, the *Pride of Carolina*, caught both tuna (hanging) and swordfish (on the deck) by setting a longline. They must have felt like they finally hit the jackpot. (Courtesy of Aycock Brown Collection, OBHC.)

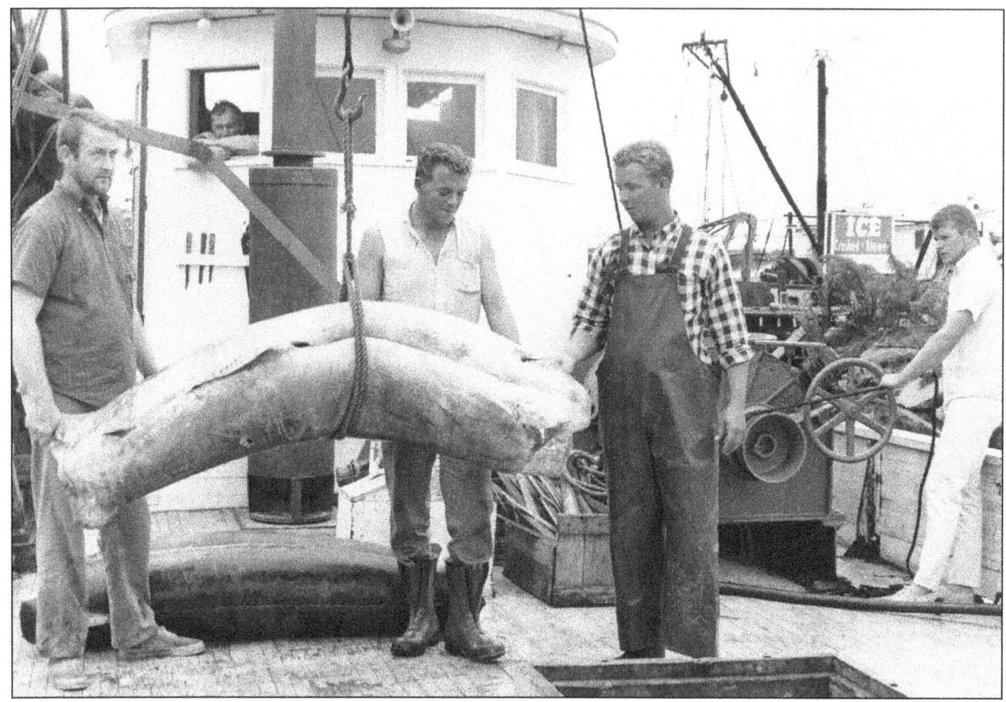

From left to right, Jack Burrus, Monty Howard O'Neal, Larry Sanderlin, Landrey Gaskins, and Butch Baum admire a monstrous swordfish. Phillip Ruhle Sr. of Baldwin, New York, began harpooning the warm-water fish in the 1940s. He studied swordfish and concluded that they were probably off of North Carolina's coast. Ruhle moved to Wanchese and liberally shared his knowledge. (Courtesy of Aycock Brown Collection, OBHC.)

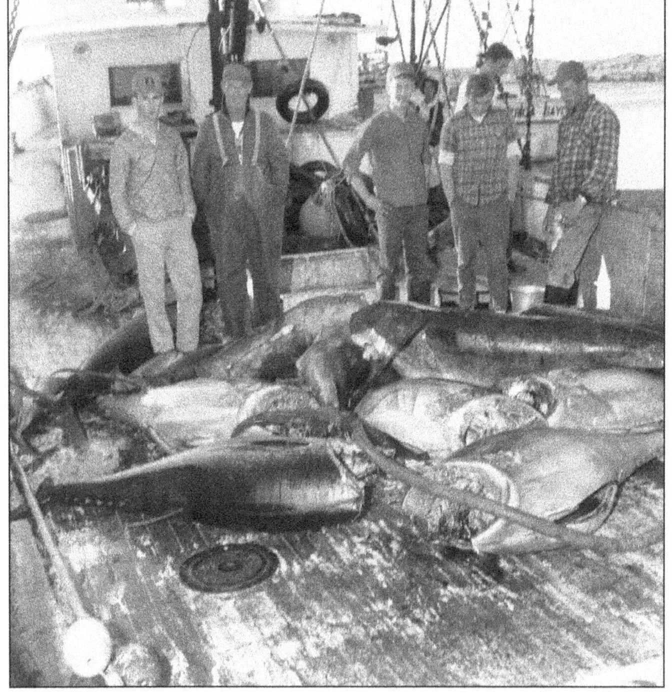

Early pioneers of swordfishing used a tremendous amount of ice for long trips at sea. A thick monofilament longline can be 20 to 80 miles long. Along the mainline at measured intervals are hanging hooks baited with mackerel. The line is suspended at a desired depth with the use of weights. A fisherman can leave his line for several hours or a few days. (Courtesy of Aycock Brown Collection, OBHC.)

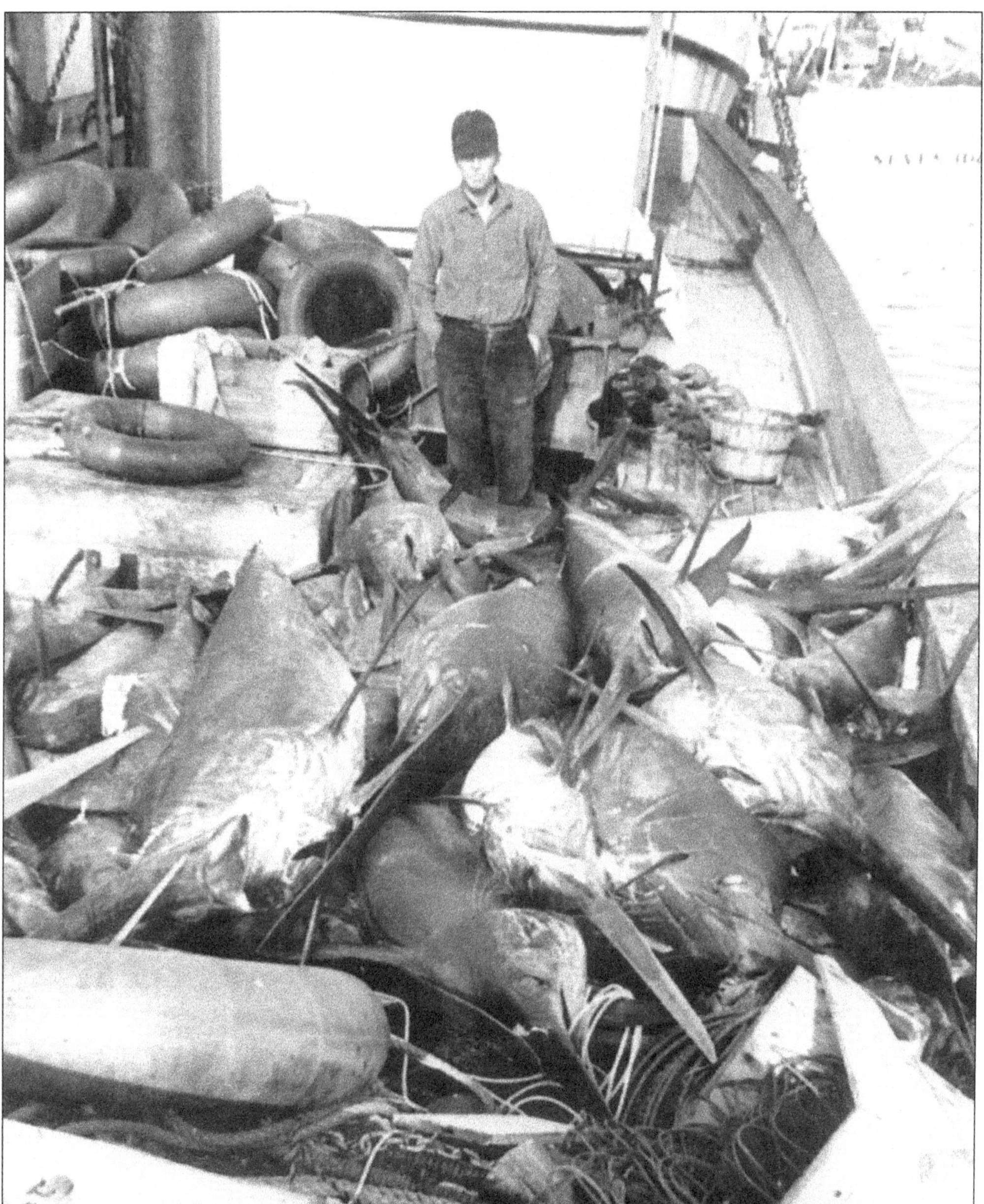

Fishing opportunities ebb and flow like the tide. In the second year that Willie Etheridge Jr. owned the *Wayne Laurin*, the usual types of fish that were normally caught were scarce. Feeling panicky about making his boat payment, he became desperate enough to try a new fishery that newcomer Phil Ruhle was doing. His first attempt at swordfishing brought in only three fish; however, on the second attempt, he hit the mother lode. Only three miles offshore, he found the tide line and set 285 baited hooks along that demarcation. Every 20 hooks, he tied on an inner tube that his crew laboriously blew up with a bicycle pump. Their efforts were rewarded with over 200 fish that all weighed over 200 pounds apiece. His catch brought in more than $10,000, a sum unheard of in 1963. The *Wayne Laurin* sank in 1974 when it was hit in the bow by a Greek freighter that kept on going. The trawler sank in just 15 minutes, but the whole crew was picked up alive and unharmed by Charles Daniels on the *Mitzi Kay*. (Courtesy of Willie Etheridge III.)

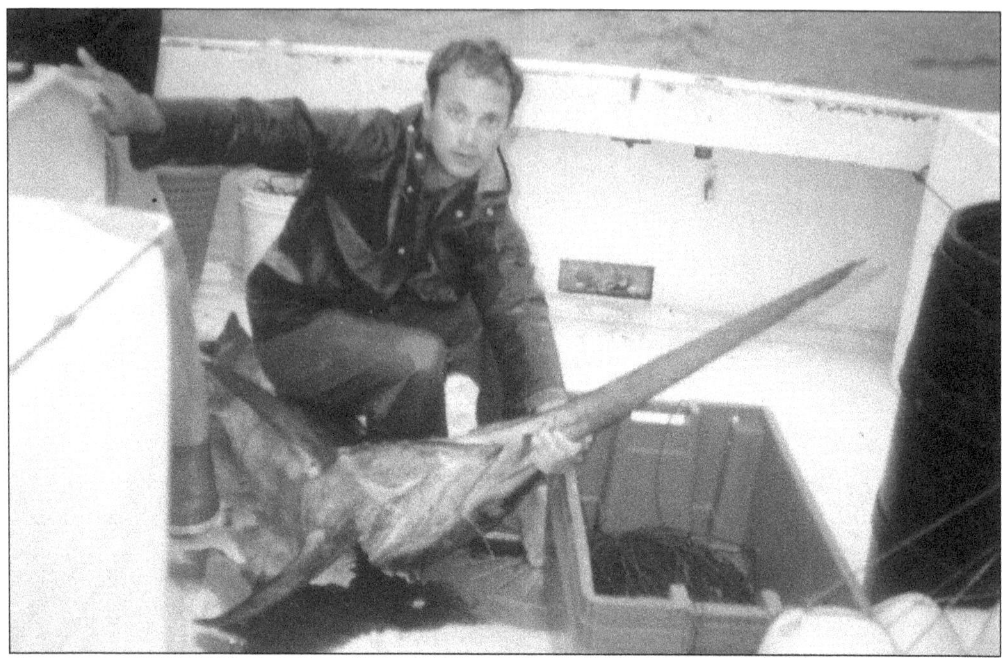

Wade Davis described his 1991 summer swordfishing with Tom Krauss as one of the best he ever had. Today, longlining boat captains are mandated to submit filmed deck activity to authorities from each trip. Sometimes, they are required to pay an observer to go along. Besides this type of fishery, only prisons are regulated in this "Big Brother is Watching You" manner. (Courtesy of Britton Shackelford.)

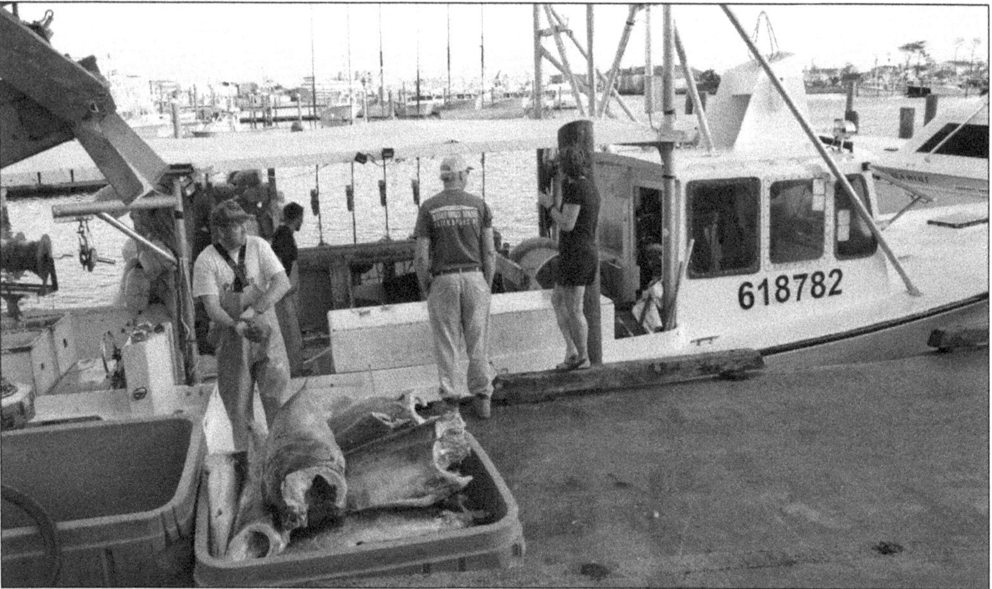

Capt. Glen Hopkins (center) was greeted by his wife, Katherine (right), after a successful February 2018 swordfishing trip. Expensive stainless-steel beeper buoys with strobe lights are lined up along the *Watersport*'s port side. Longlines are attached to them in hopes of insuring no loss of gear. Hopkins has used the Internet as a way of voicing his concerns over unfair fishing regulations. (Courtesy RWG.)

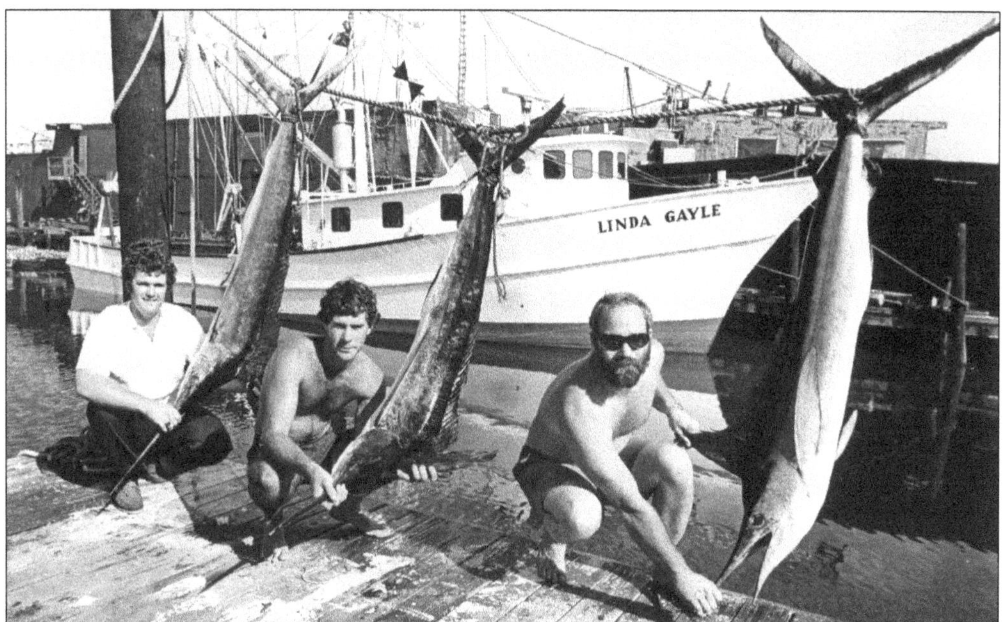

From left to right, Billy Carl Tillett, Steven Etheridge, and John Arendts hold up the bills of white marlins they caught while scalloping in 1978. During the long trip home, they crossed the ocean canyons off of Maryland and decided to try their luck. They were always looking for a little entertainment after they had caught all the scallops the *Linda Gayle* could hold. (Courtesy of Billy Carl Tillett, photograph by J. Foster Scott.)

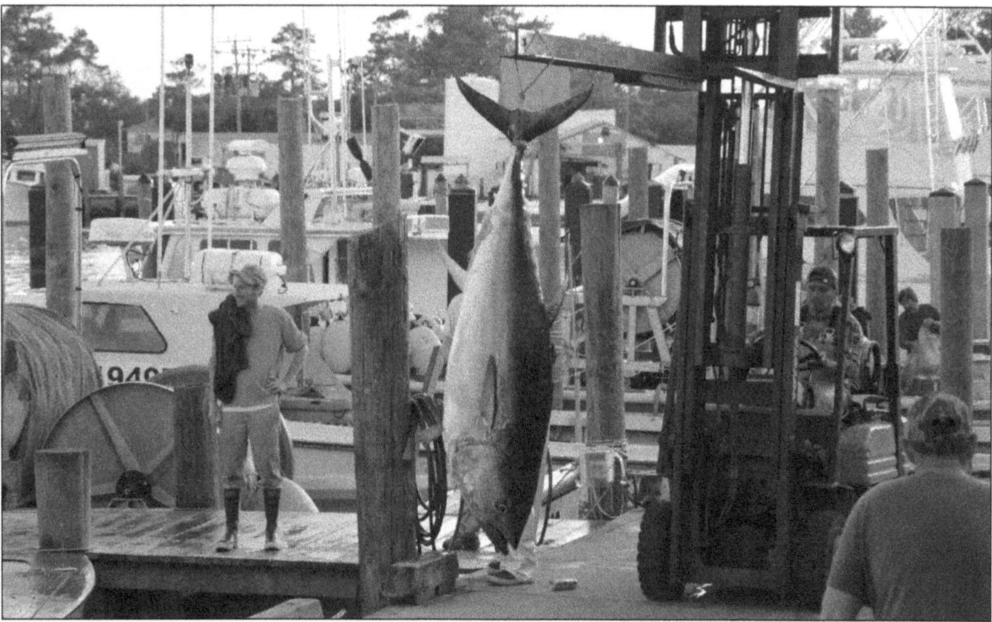

From the moment an Atlantic bluefin tuna is landed, extreme care is used in every aspect of its handling. A forklift gently raised one from a boat's deck in February 2018. As the most expensive fish in the world, they have to be in perfect condition to get high prices. Bluefin are prized for their fat content, which makes them the top choice among sushi and sashimi consumers. (Courtesy of RWG.)

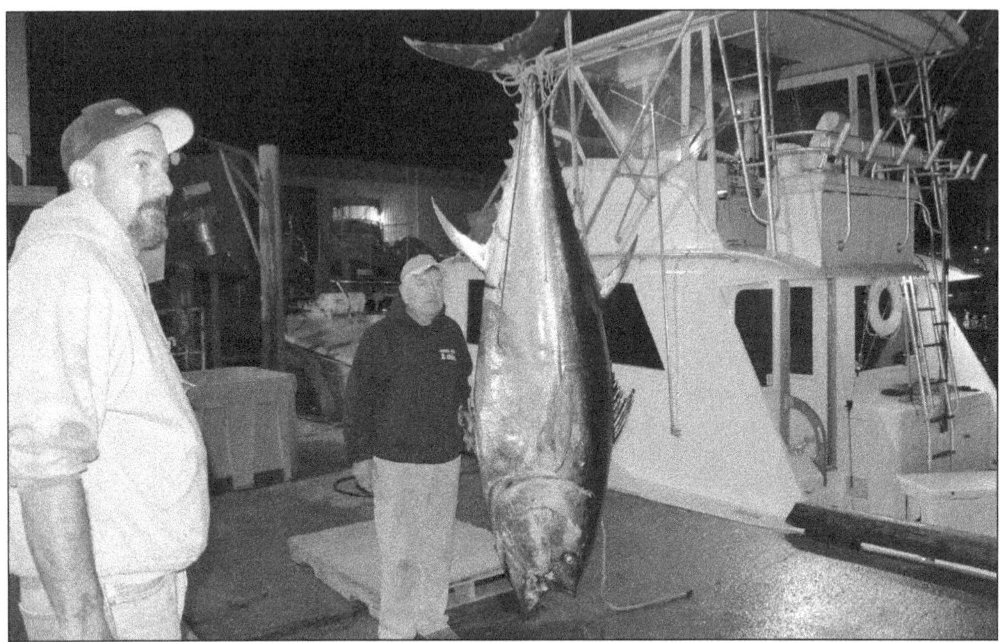
Jason Peyton (left), working for tips, waited for the next captain to ask him to clean a tuna. A federally imposed quota on bluefin tuna allows charter and head boat anglers, as well as commercial fishermen, to take part in the harvest. Fish dealers have to report landings within 24 hours of receiving a bluefin to keep the tonnage totals up-to-date. (Courtesy of RWG.)

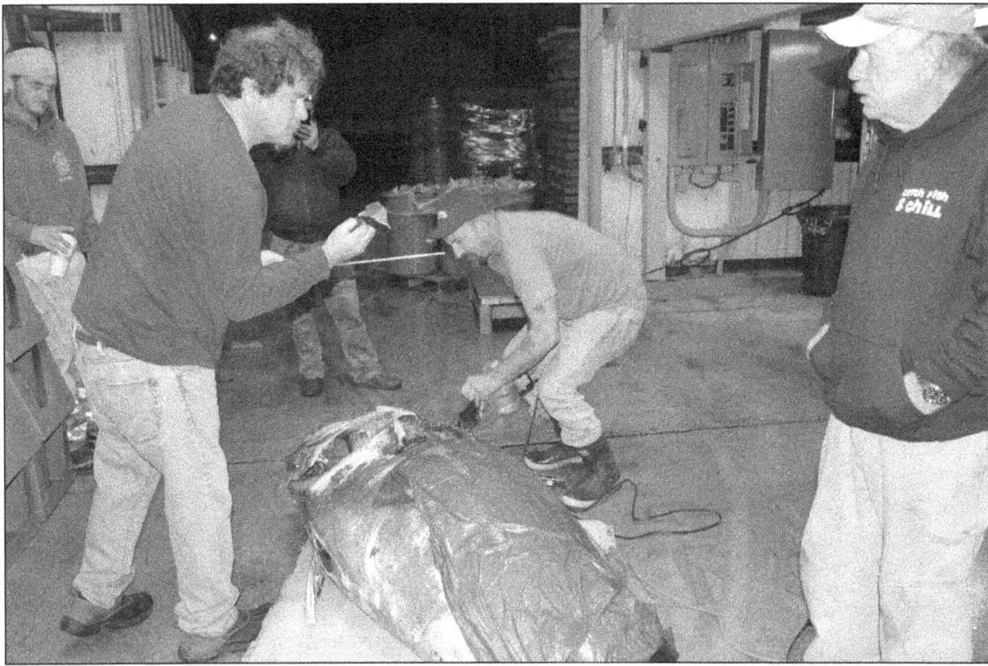
A tuna buyer from New Hampshire inspected a sample steak from a 400-plus-pound bluefin as Peyton finished dressing it out. The buyer will decide to which market the fish will be transported, with the best sushi-grade tunas being flown to Japan. Tuna season can be exhausting for everyone as fishermen race to catch what they can before the quota is met. (Courtesy of RWG.)

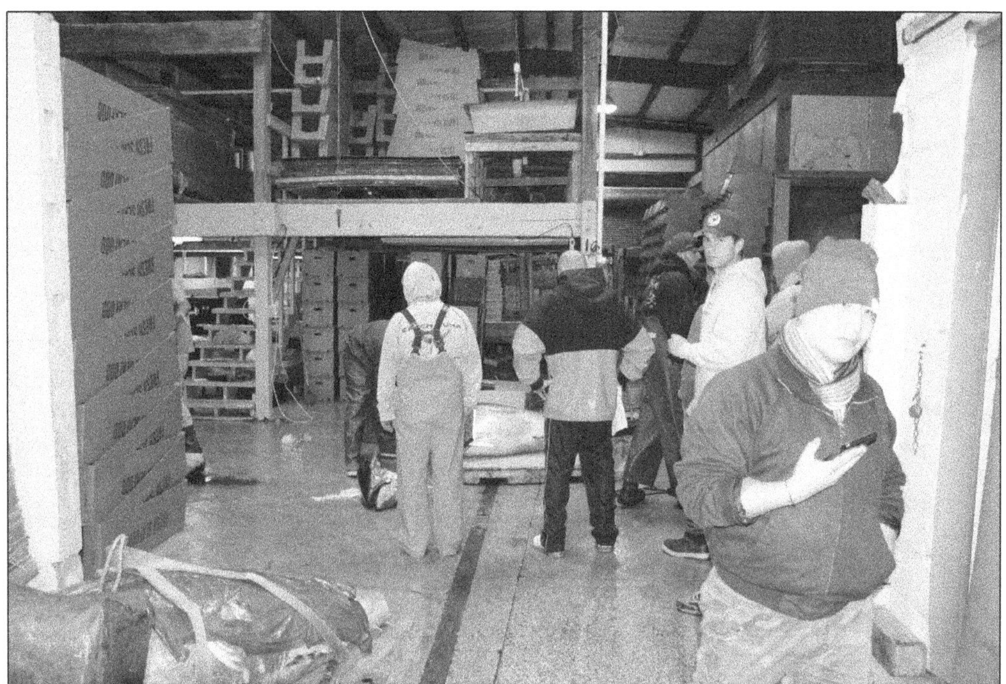

Talking nonstop on the phone except when he was taking and transmitting photographs, an Asian broker paced the floor at O'Neal's looking for the best buy. There is a great disparity between what policy makers calculate fish populations to be and what fishermen report them to be. One side says they are in decline, and the other says they are in abundance. (Courtesy of RWG.)

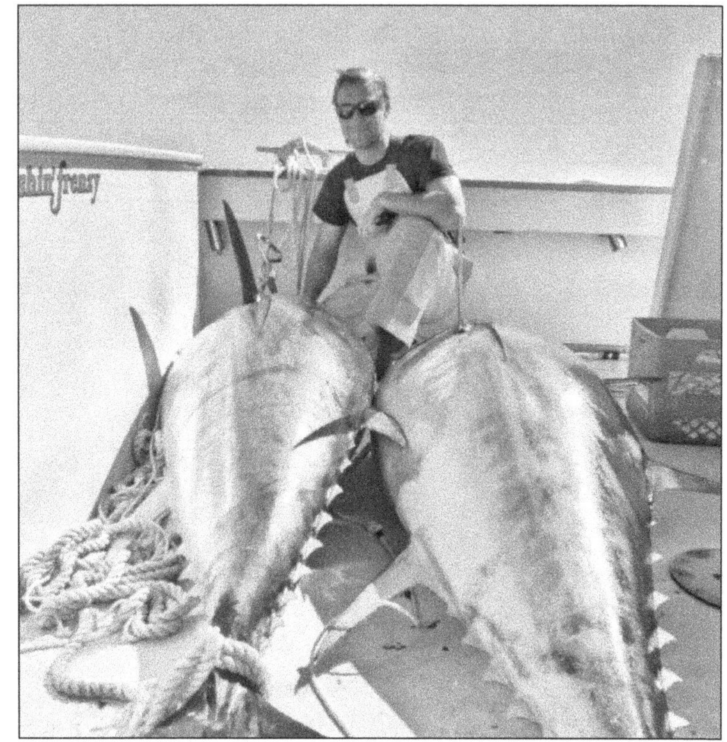

A cast member of National Geographic Channel's *Wicked Tuna: Outer Banks*, Billon Hollingsworth, was upstaged by the superstars of the show. Hollingsworth has been working as a mate since high school. Many watermen are discouraging their sons from going into their line of work because the future of fishing seems so grim. It would be very hard to make it without first being established. (Courtesy of Billon Hollingsworth.)

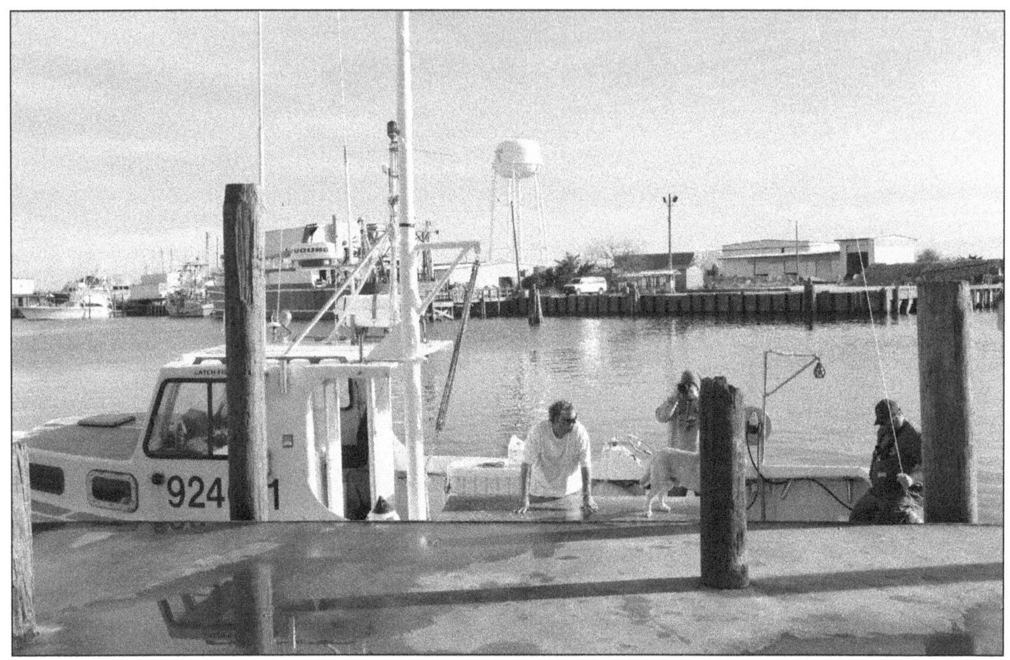

A 2018 production crew from *Wicked Tuna* filmed as Capt. Charlie Griffith on his boat *Reels of Fortune* brought a tuna to the dock. Local fish dealers have refused to be part of the program, stating that higher-than-retail prices broadcast by the show are fictitious and do not help the plight of the average hardworking fisherman. (Courtesy of RWG.)

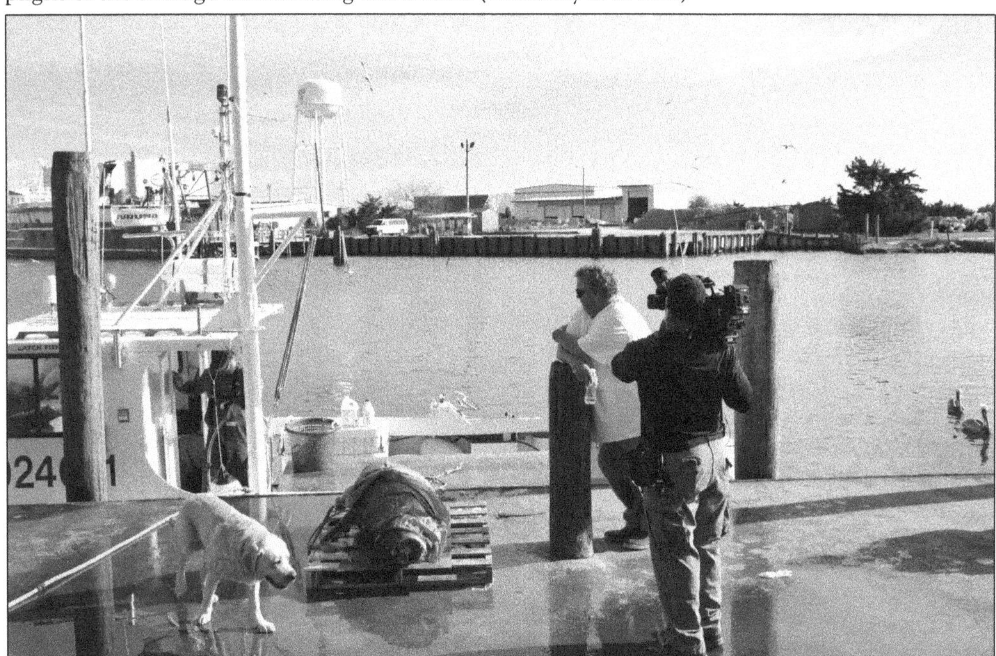

The yellow lab Renegade skirted around his master's catch. The tuna's name is derived from a Greek word meaning "I rush, dart along," since it hunts in groups that herd schools of small fish by zigzagging. Tuna spend the winter off of Hatteras Island in the warm waters of the Gulf Stream. Their great strength, stamina, speed, and size make them a challenge to bring in. (Courtesy of RWG.)

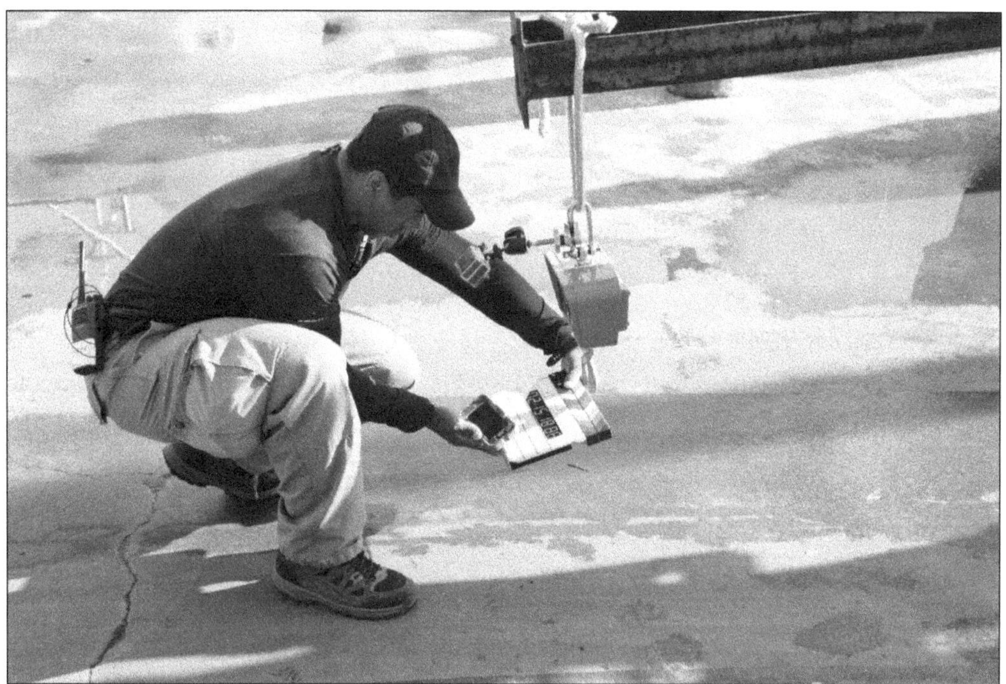

Here, a *Wicked Tuna* film crewmember uses an electronic version of the slate movie clapper. Fishermen have wished they could retake scenes in their journey with governmental superintendence. The majority of fishermen feel bulldozed by a process that has been political, illogical, and one-sided. They consider themselves to be abandoned by those who are supposed to represent them. (Courtesy of RWG.)

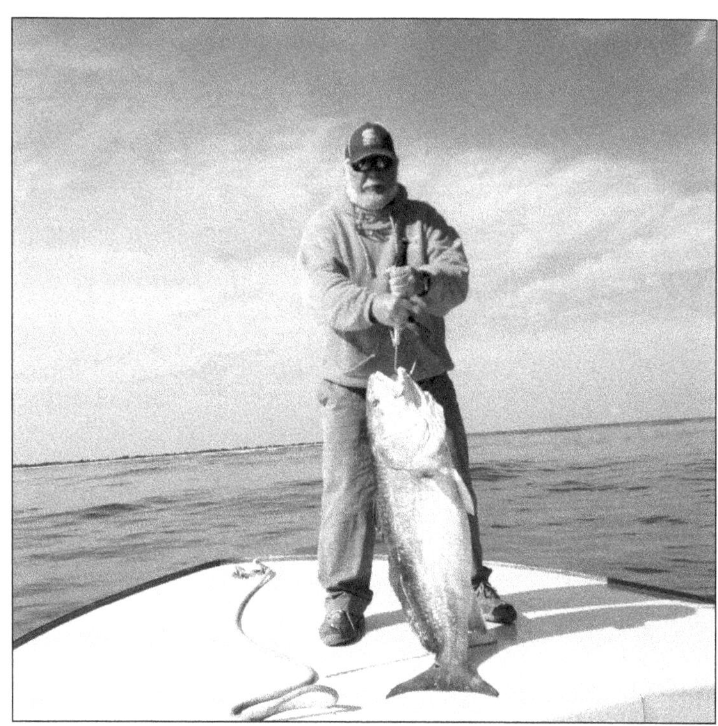

Unlike his father, Willie Etheridge III had no love affair with boats and got seasick whenever he went on the water. He took over his grandfather's fish business in the 1970s after returning from Vietnam. The Willie R. Etheridge Seafood Company became a powerhouse, often packing 20,000 pounds of flounder a day. His once tireless advocacy for fishing has become dampened by frustration. (Courtesy of Skip Dixon.)

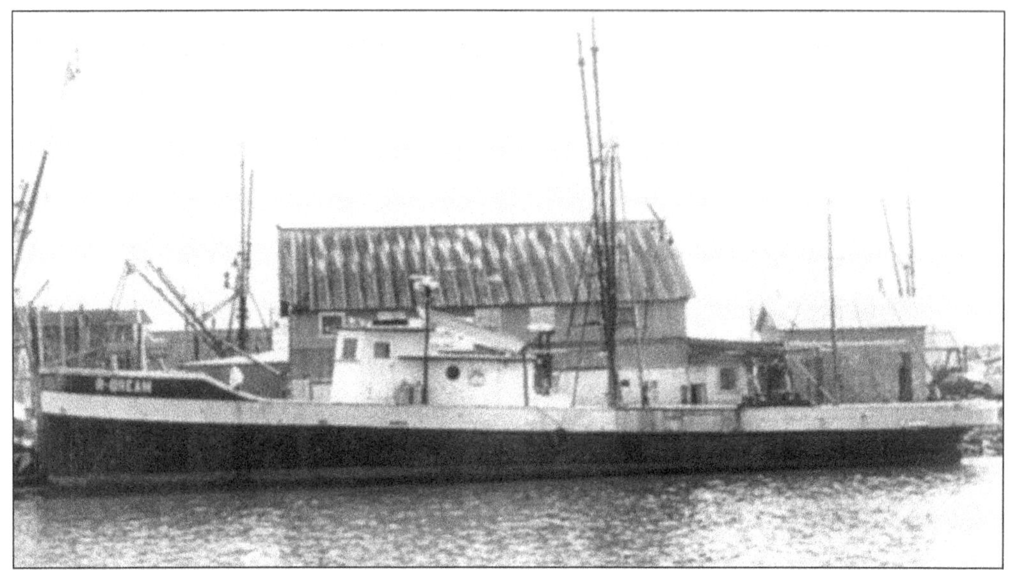

Plying the waters for a living may not be an achievable dream for a waterman much longer. Commercial fishermen desire to be part of the management process. So far, they have seen their own state and national fisheries commissions made up of people who have no connection whatsoever to fishing. Decisions are made without the ones who know the resources most intimately. (Courtesy of Mark Vrablic.)

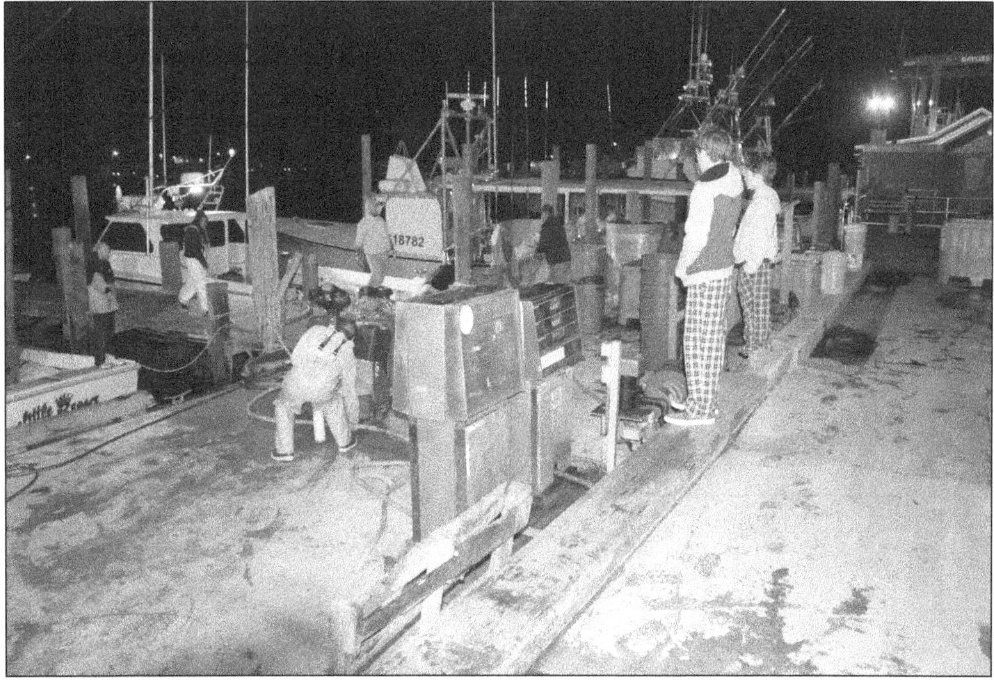

Before going to bed, Caleb and Braden O'Neal had to go down to the dock one last time to see what was being brought ashore. Many environmental and conservation groups have adopted a hard line and will not cease their political lobbying until all commercial fishing is stopped. It might make a difference in commissioners' voting if they could only see the bright faces of these future watermen. (Courtesy of RWG.)

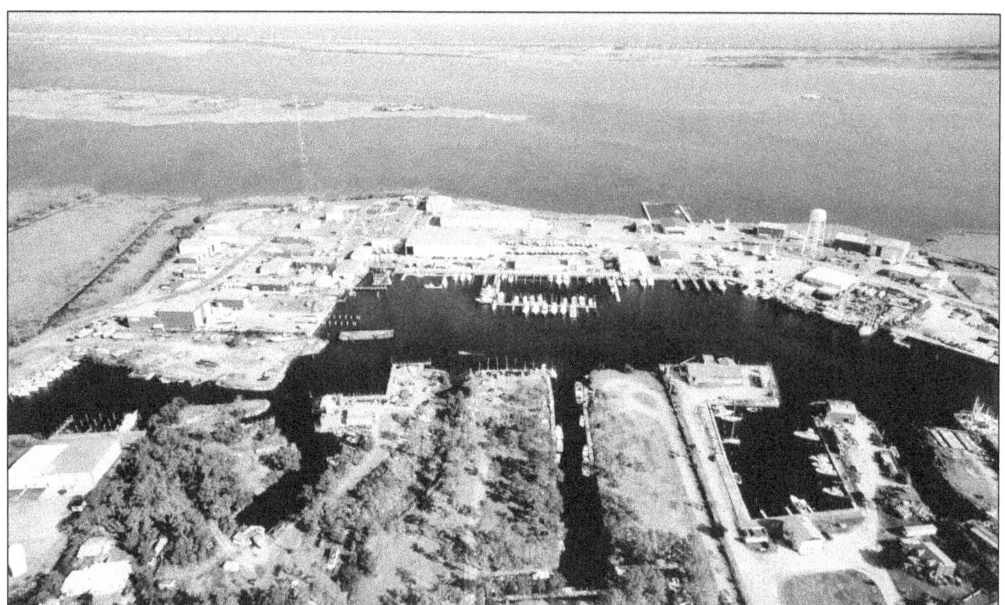

What high hopes local and state officials had when the Wanchese Seafood Industrial Park was officially completed in March 1981. The park was developed as a result of a federal promise to permanently stabilize Oregon Inlet. As that dream faded, boatbuilding and repair facilities began to dominate the recently renamed Marine Industrial Park. (Courtesy of Alvah Ward, photograph by Ray Matthews.)

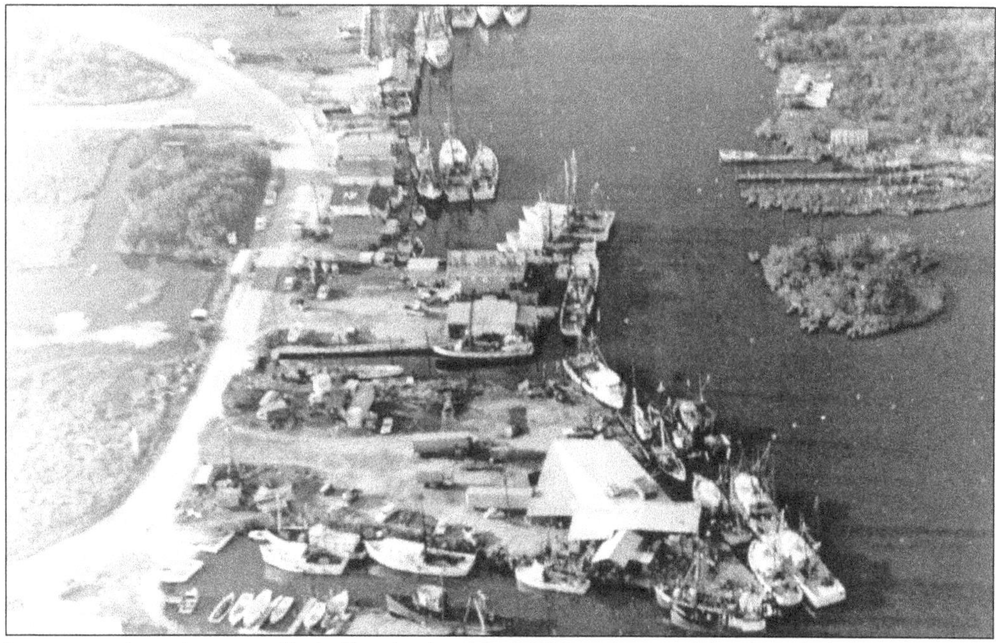

It is astounding that a small, shallow harbor like Mill Landing in Wanchese led the country for many years in packing scallops, flounder, swordfish, and sharks. Irrepressible Outer Bankers have always faced obstacles just by living on barrier islands. Commercial fishermen want to overcome adversity and experience the freedom of continuing to provide wild caught, affordable, fresh seafood to the masses. (Courtesy of Mark Vrablic.)

www.ingramcontent.com/pod-product-compliance
Lightning Source LLC
LaVergne TN
LVHW060624110826
845147LV00015B/935

* 9 7 8 1 4 6 7 1 0 3 3 5 0 *